Springer Theses

Recognizing Outstanding Ph.D. Research

For further volumes:
http://www.springer.com/series/8790

Aims and Scope

The series "Springer Theses" brings together a selection of the very best Ph.D. theses from around the world and across the physical sciences. Nominated and endorsed by two recognized specialists, each published volume has been selected for its scientific excellence and the high impact of its contents for the pertinent field of research. For greater accessibility to non-specialists, the published versions include an extended introduction, as well as a foreword by the student's supervisor explaining the special relevance of the work for the field. As a whole, the series will provide a valuable resource both for newcomers to the research fields described, and for other scientists seeking detailed background information on special questions. Finally, it provides an accredited documentation of the valuable contributions made by today's younger generation of scientists.

Theses are accepted into the series by invited nomination only and must fulfill all of the following criteria

- They must be written in good English.
- The topic should fall within the confines of Chemistry, Physics, Earth Sciences, Engineering and related interdisciplinary fields such as Materials, Nanoscience, Chemical Engineering, Complex Systems and Biophysics.
- The work reported in the thesis must represent a significant scientific advance.
- If the thesis includes previously published material, permission to reproduce this must be gained from the respective copyright holder.
- They must have been examined and passed during the 12 months prior to nomination.
- Each thesis should include a foreword by the supervisor outlining the significance of its content.
- The theses should have a clearly defined structure including an introduction accessible to scientists not expert in that particular field.

Michail Bachtis

Heavy Neutral Particle Decays to Tau Pairs

Detected with CMS in Proton Collisions at $\sqrt{s} = 7$ TeV

Doctoral Thesis accepted by
University of Wisconsin–Madison, USA

Springer

Author
Dr. Michail Bachtis
Department of Physics
University of Wisconsin–Madison
Madison, WI
USA

Supervisor
Prof. Sridhara Dasu
Department of Physics
University of Wisconsin–Madison
Madison, WI
USA

ISSN 2190-5053 ISSN 2190-5061 (electronic)
ISBN 978-3-319-38078-0 ISBN 978-3-319-03257-3 (eBook)
DOI 10.1007/978-3-319-03257-3
Springer Cham Heidelberg New York Dordrecht London

Printed on acid-free paper

Springer is part of Springer Science+Business Media (www.springer.com)

To all those who continuously seek the truth…

Supervisor's Foreword

The Ph.D. thesis of Michalis Bachtis is an ideal report of his work in experimental particle physics, which ranged from contribution to the trigger system of the experiment to final results in search for the Higgs boson. Such breadth is quite rare in recent theses in particle physics, and as such this work stands out. The thesis covers development of hardware and software triggers, reconstruction techniques of hadronic τ decays, methods for estimating and reducing backgrounds for τ pair events, measurement of DY cross section at Z-pole using $Z \rightarrow \tau^+\tau^-$ events, estimation of τ reconstruction efficiency, and ultimately search for $H \rightarrow \tau^+\tau^-$ decays. The unique τ_h reconstruction algorithm, which reconstructs the decay modes has effectively raised the standard τ_h reconstruction in hadron colliders to a new level. Therefore, this body of work has become the basis for the CMS experimental searches of not only Higgs boson(s) but also other new particles that decay in τ modes using Bachtis's algorithm. The τ-mode Higgs searches using CMS have become several factors better than originally estimated during the preparation of the CMS experiment. It gives me immense pleasure to realize that Bachtis's work was selected for the 2012 CMS thesis award and is now being published in the Springer theses series.

Madison, WI, September 2013 Prof. Sridhara Dasu

Abstract

This thesis describes a study of the tau pair final state in proton collisions at a center of mass energy of 7 TeV using the CMS Experiment at the CERN Large Hadron Collider. Tau leptons usually decay to hadrons and their efficient detection is very challenging in a hadronic environment. A novel tau reconstruction algorithm that maintains tau detection efficiency for low background rates was developed. Events were then selected from the CMS 2011 data sample in three $\tau\tau$ final states depending on the tau decays: $\mu + \tau_h$, $e + \tau_h$, $e + \mu$ where τ_h denotes a hadronic tau decay. Those events were used to measure the Z-boson production cross section where $Z \rightarrow \tau\tau$. A search for the Standard Model and Minimal Supersymmetric Standard Model (MSSM) Higgs bosons was also performed. No signal was observed in the Higgs search and stringent new bounds were set for the SM Higgs production and in the MSSM parameter space.

Acknowledgments

If you asked me 15 years ago, what is the probability that I would get a Ph.D., the answer would have been zero simply because I did not even know what a Ph.D. was back then. Therefore, I feel I should acknowledge all the people who helped me make this happen. First of all my parents, Sotiris and Adamantia who brought me into this world and raised me with love and care. It is remarkable that they pushed me toward science since I was very young, without even having completed high-school education themselves. Second, my little brother Jimis who managed to tolerate me all those years.

Life in the village of Agios Nikolaos Vion would not be that pleasant without some friendships that began since I was a child and still last. I would like to thank my friends Dimitris A, Giannis D, and Giannis K for the great times we had together that are still repeated when we meet. I want to thank especially Giannis K because he taught me to be strong and not give up by being strong himself after his terrible motorcycle accident.

During my studies in the National Technical University of Athens, I met some great people and especially my advisors Theo Alexopoulos and Yorgos Tsipolitis. I would have never become a particle physicist if I had not met them. I will never forget Yorgos teaching me how to produce robust results without errors. Also, working with Theo all night in the office and lab trying crazy ideas that, yes, worked. I want to thank both of them for recommending me for doctoral studies. During the same period while working in ATLAS experiment I met Sotiris Vlachos. Sotiris taught me a lot about detector physics and statistics in particle physics. Moreover he has been like a big brother to me since I moved to CERN.

During my 7-year stay in Athens I made some very good friends. I especially want to thank Panos for being more than a brother to me. We had wonderful road trips on motorcycles and luckily we are both still alive.

After some time in Geneva I moved to Wisconsin in the summer of 2007, where I met professors Sridhara Dasu and Wesley Smith. Sridhara is an excellent advisor. I learned a lot from him about physics, data analysis, trigger and computing, and mostly he has been very supportive in all of my steps during the completion of this dissertation being first a friend and then my boss. Wesley transmitted me his great enthusiasm about physics and technology and provided me huge amounts of knowledge about trigger, detector, and physics. I want to thank both of them for giving me independence during my research and for allowing me to follow my own research path in the CMS experiment.

Life in Madison would not have been so nice If I had not met my friends Alex and Lisa. We had a great time and they helped me adapt to the different lifestyle. At the same time, I want to thank Krista for her infinite love and support and for making my life in Madison so beautiful.

Back in Geneva, working on CMS this time, I had the chance to work with some wonderful people. I want to thank Sascha Savin for all the help and knowledge he gave me about data analysis, Sasha Nikitenko for being a mentor in Higgs and tau physics, and Monica Vazquez Acosta for the great times during the $Z \rightarrow \tau\tau$ work. I also want to thank Simone Gennai for the excellent times we had working on tau trigger and identification and also for the fun climbing sessions. Also, Markus Klute for all the things I learned from him while we published the $H \rightarrow \tau\tau$ results five times in 2011. I would also like to thank the Higgs group conveners Vivek Sharma and Christoph Paus for the excellent communication and collaboration during the time this thesis work was performed and also CMS physics coordinator Gigi Rolandi and spokesperson Guido Tonelli for their strong support to the tau community in CMS.

Finally, I would like to thank my colleagues in Geneva: Christos, the other part of the Wisconsin Greek doublet, for being an excellent friend and officemate and for helping me in bad times like during my motorcycle accident. Also my officemates Kira, Marc, Josh, and Ian for managing to stand being in the same office with me for so long. I want to thank my friend Loukas for being so close to me during the last years despite the short period we knew each other and my friend Antonis for the nice times we had in Geneva. Finally, I want to thank my colleague and friend Isobel for helping me to become a better teacher by learning from me, for being supportive during the hard period of my thesis writing, and for managing to help me to find my old self and pull myself out of the office and the terminal sometimes.

Contents

Chapter 1
The Standard Model of Particle Physics

Pursuit of answers to fundamental questions of life and nature is time invariant. The atomic hypothesis of matter was first brought up by Democritus in 400 B.C. Democritus stated that everything consists of atoms which are physically but not geometrically indivisible. It took 2000 years to reach the era of Galileo Galilei and Isaac Newton to move from pure philosophy to scientific methodology as known today. This chapter describes the Standard Model of Particle Physics starting from a historical evolution.

1.1 Historical Approach

The twentieth century can be characterized as the golden age of physics. This paragraph describes a summary of the historical evolution of particle physics. A more detailed description can be found elsewhere [1, 2]. The era of particle physics essentially started with the discovery of the electron(e) by J. J. Thomson in 1897 by noticing that cathode rays get deflected by magnetic fields, leading him to the conclusion that the rays consist of charged particles, the electrons. Thomson believed that the electrons were basic constituents of the atoms. To explain the neutral charge and much higher mass of the atom compared to the electron he assumed that electrons were embedded within a heavy positively charged paste. At the same time, studies of black-body radiation by Planck in 1900 and the photoelectric effect explanation by Einstein in 1905 showed the particle nature of the photon. Thomson's hypothesis was repudiated by Rutherford's scattering experiment. Rutherford showed in 1911 that the mass and the charge is concentrated in the center of the atom. Both Thomson's and Rutherford's observations led Niels Bohr in 1914 to propose the model of hydrogen atom with one proton(p) and one electron orbiting around it. Bohr's model showed spectacular agreement with experiment also for larger atoms. The difference between the mass of the larger atoms with the one predicted by Bohr was explained later by the discovery of the neutron(n) by Chadwick in 1932.

M. Bachtis, *Heavy Neutral Particle Decays to Tau Pairs*,
Springer Theses, DOI: 10.1007/978-3-319-03257-3_1,

The era of non-relativistic quantum mechanics ends with the proposal of Dirac's equation. Einstein's special theory of relativity was introduced since 1905 but took about 20 years to be applied by Dirac in his famous equation that combines the high momentum physics described by Einstein's relativity and that of smallest distances described by quantum mechanics. Although Dirac's equation appeared to be correct, it had negative energy solutions. Those could be explained by the presence of a particle that has the same mass as the electron but with positive charge. The triumph of the theory was established with the discovery of the positron by Anderson in 1932. Dirac gave many interpretations of the negative energy solutions but in 1940s Feynman and Stueckelberg interpreted negative energy as positive energies of a different particle. This introduced the notion of the antiparticles which is a fundamental aspect of quantum field theory. Quantum electrodynamics(QED) was developed by Feynman and Schwinger to consistently describe electromagnetic interactions between charged particles with photon exchange mediating such interactions. QED remains the most well tested scientific theory ever.

One unanswered question at the time was what binds the nucleus together? Yukawa proposed a new field, and a force carrier particle, the meson, that had not been observed in the laboratory. Cosmic ray experiments led by Anderson, Neddermeyer, Street and Stevenson identified particles matching Yukawa's description in cosmic rays. However, other experiments in Rome showed that the cosmic ray particles interact very weakly with nuclei which is against the Yukawa hypothesis, that those particles are the force carriers that strongly bind the protons. It took 10 years for Powell to prove that there are two particles in cosmic rays, Yukawa's pion that disintegrates to a weakly interacting lighter charged particle, the muon.

Another astonishing observation in the same period was the spectrum of electrons produced in nuclear β decay. Assuming that one nucleus disintegrates to a lighter nucleus and an electron, the energy of the electron can be explicitly predicted by the kinematics. However, it was shown that the experimental energy spectrum of the electrons was softer and broader than the energies predicted. This effect was explained by Pauli by introducing one additional particle in the final state, the neutrino(ν). At the same time, Powell's experiments on the cosmic rays showed that the pions in the atmosphere that disintegrate into muons are produced at angles with respect to the original particles showing the presence of an additional particle in the pion decay. That was believed to be Pauli's neutrino. Powell went further and established that the muon also decays to electron but with two neutrinos in the final state this time.

The neutrino was established experimentally in the 1950s by Cowan and Reines. They studied the inverse beta decay process $\bar{\nu}+p^{+} \rightarrow n+e^{-}$ in a large tank of water in a nuclear reactor, and by identifying the positrons they gave unambiguous evidence for the neutrino existence. Another peculiarity regarding the neutrinos is that the decay $\mu \rightarrow e+\gamma$ was never observed. This shows that there is different lepton family number conservation, i.e. conservation of the μ family number and conservation of e family number. For this conservation to hold, the subsequent neutrinos should be different for each family; there should be muon-neutrinos and electron neutrinos. Under this assumption, the pion decay would become: $\pi^{-} \rightarrow \mu^{-} + \bar{\nu}_{\mu}$ while the muon decay becomes: $\mu^{-} \rightarrow e^{-} + \bar{\nu}_{e} + \nu_{\mu}$. The test of this hypothesis was studied in Brookhaven by Lederman, Schwartz and Steinberger. They proved that while the reaction $\bar{\nu}_\mu + p^+ \rightarrow \mu^+ + n$ is allowed, the reaction $\bar{\nu}_\mu + p^+ \rightarrow e^+ + n$ is forbidden. The situation started to become much more complicated when mesons started to be observed. Rochester and Butler observed a neutral particle decaying into two pions ($K^0 \rightarrow \pi^+ + \pi^-$) and named it the Kaon while Powell observed a charged Kaon decaying to three pions ($K^+ \rightarrow \pi^+ + \pi^+ + \pi^-$). In the following years many more mesons were discovered such as the ω , the ρ and others. Meanwhile, the Lambda particle, a heavier strongly interacting particle, baryon, was discovered in 1947 by Anderson's group at Caltech via the decay $\Lambda \rightarrow p^+ + \pi^-$. This discovery was very important because it extended the baryon family. The conservation of baryon family number was introduced earlier by Stueckelberg to ensure the stability of the proton. Since the Λ decay was observed and there is a proton in the final state, Λ must belong to the baryon family. Another observation regarding mesons and baryons was that they were produced very often but decay relatively very slowly. Pais and others the deduced that mesons and baryons are produced via strong interaction but decay via weak interactions. To account for this Gell-Mann and Nishijima introduced a new quantum number, the strangeness that is conserved for strong interactions but was not conserved in weak interactions. With the discovery of several mesons and baryons, the list of particles became large. Gell-Mann introduced the Eightfold Way which arranged the particles based on their charge and strangeness. While organizing the baryons, Gell-Mann predicted a new particle with strangeness of -3 and negative charge, the Ω^-. This particle was soon observed and the Eightfold Way was proven.

The Eightfold Way was the periodic table of elements of particle physics but this was only the beginning of modern particle physics history. Gell-Mann and Zweig proposed in 1964 that mesons and baryons have structure and consist of fundamental particles called quarks. They proposed three quarks, u, d and s with charge 2/3, $-1/3$ and $-1/3$ and strangeness 0, 0 and -1 respectively. Under this model, every baryon consists of three quarks and every meson consists of a quark–antiquark pair. The weakness of the quark model was that no quark was observed experimentally. In addition, by the Pauli exclusion principle three identical quarks in a baryon cannot be in the same state. This required that quarks are subject to one more quantum number, color that distinguishes the identical quarks in such baryons, solving the Pauli exclusion principle issue. In addition, requiring that all physical final states must be colorless allows bound states of baryons and mesons (two or three quark pairs). Furthermore it does not allow free quarks to exist in nature. Evidence for the proton substructure in terms of quarks (partons) was obtained by deep inelastic scattering experiments that probe the structure of hadrons following the same principle as Rutherford's experiment.

In 1970 Glashow et al. [3] categorized particles in generations so that there is symmetry between the leptons ($(e\nu_e)$, (μ, ν_μ)) and quarks ((u, d), (s, c)). They introduced a new quantum number, weak isospin that needs to be conserved in weak interactions. However the proposed GIM mechanism predicted an additional quark c that was not observed yet. At the same time, the J/ψ meson was discovered by C. C. Ting's group at Brookhaven and B. B. Richter's group at SLAC in 1974. The J/ψ particle was heavier than the proton and had a very long lifetime compared to the other mesons. The explanation for the J/ψ particle was that it should be a bound state of two charm quarks ($c\bar{c}$) validating the GIM mechanism. The generational structure was extended in 1975 with the discovery of the τ lepton by M. Perl. A third quark generation was added by the discovery of the b quark with the observation of the Υ meson which is a $b\bar{b}$ bound state. The t quark was observed in Fermilab's Tevatron in 1995 by observing top pair production.

Discovery of meson and baryon substructure required a new fundamental theory of strong nuclear forces to replace Yukawa's meson mediation theory. One new particle, the gluon was proposed to mediate the strong interactions. The gluon must couple to the quarks therefore it must carry color. Gluons exists in eight color configurations. It was discovered in DESY in late 1970s by studying events with three hadron showers (jets) in the final state. Via the reaction $e^+e^- \rightarrow q\bar{q}$, due to color confinement, the two quarks appear as a stream (jet) of hadrons in the final state. Those events are expected to have two of those jets of hadrons in the final state. If the gluon exists, in a similar way to the radiation of a photon, a quark could radiate a gluon resulting in a reaction of the form $e^+e^- \rightarrow q\bar{q}g$. This signature would result in three jets of particles. This was observed and was an indisputable evidence of the presence of the gluon.

Fermi had treated the weak interactions in terms of a contact theory requiring no mediating particle. Fermi's theory was proven to be very effective at low energies, however it was bound to fail at higher energies. Theorists believed that to solve this problem, new massive ($\mathcal{O}(100)$ GeV) electrically charged mediating particles had to be introduced. Glashow [4], Weinberg [5] and Salam [6] proposed the electroweak theory which unified electrodynamics and weak interactions with the addition of a neutral Z boson and was confirmed by the observation of neutral currents at CERN in 1973. The massive mediating bosons were discovered later by Carlo Rubbia's group in 1983 with the UA1 experiment at the CERN SPS accelerator.

One problem of the electroweak theory was that W and Z bosons had to be massless for the theory to be consistent while for the weak interactions to be weak they are very heavy. Furthermore, fermions are required to be massless in the theory. The mass of the vector bosons was explained by the spontaneous symmetry breaking of weak interactions via the Higgs mechanism proposed by Brout and Englert [7], Guralnik et al. [8] and Higgs [9–11]. The Higgs mechanism introduces a scalar doublet consisting of four degrees of freedom. Three of those degrees of freedom are absorbed to give mass to the W and Z bosons. The remaining degree of freedom appears as a new particle the Higgs boson that has not been observed yet. The search for the Higgs boson is the main subject of this thesis.

1.2 Quarks, Leptons and Gauge Bosons

Summarizing the historical introduction, the Standard Model of Particle Physics consists of six quarks and six leptons grouped in three generations. Higher generation particles decay via weak interactions to lower generation particles, explaining why the low energy world as it is known today consists only of particles of the first generation.

All matter particles are fermions having spin of 1/2. The quarks (u, c, t) have a charge of $+2/3$ while the quarks (d, s, b) have a charge of $-1/3$. The three leptons e^-, μ^-, τ^- have charge of -1 while the corresponding neutrinos are neutral. Neutrinos were introduced in the SM as massless particles however recent results from neutrino oscillation experiments [12–14] point to neutrinos having very small but non-zero masses. The anti-particles of all 12 particles exist and have opposite charge. Each fundamental force is associated with spin 1 mediator particles. The strong force is mediated by eight colored gluons, and the electromagnetic force is mediated by the photon, while the weak interactions are mediated by the $W^{\pm}$ and Z bosons. Gravity is extremely weak to have visible effects in the TeV scale. Figure 1.1 shows the SM particles in each generation as well as the force mediators. However, there is one missing piece of the Standard Model the Higgs boson.

The Standard Model is formulated mathematically as a field theory under the $SU(3) \times SU(2)_L \times U(1)$ symmetry. $SU(3)$ describes color and interactions between gluons and quarks while $SU(2)_L \times U(1)$ describes electroweak interactions. By Noether's theorem each symmetry corresponds to a conservation law yielding a conserved quantity, which, in the case of the Standard Model, are color, weak isospin and hypercharge. In the unbroken $SU(2)_L \times U(1)$ symmetry, gauge bosons and

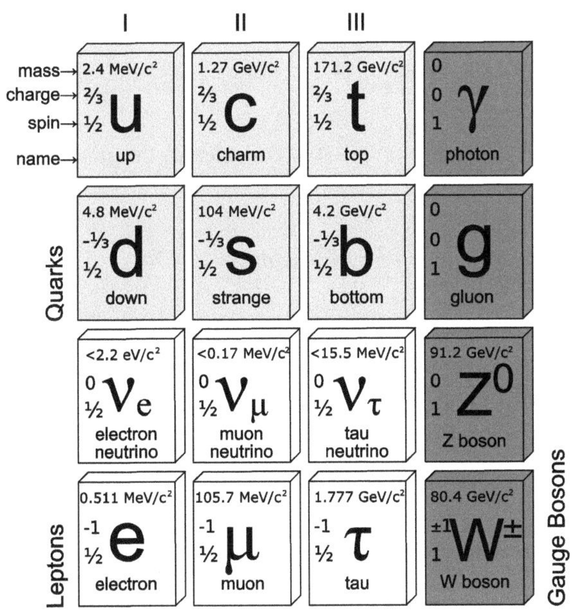

Fig. 1.1 Chart of the standard model fermions and gauge bosons grouped in generations

fermions are massless. The mass of the weak bosons $W^{\pm}$, Z is explained by breaking of the symmetry between the electromagnetic and weak interactions via the Higgs mechanism described in the next section.

1.3 The Higgs Mechanism

The price of electroweak unification is that the gauge bosons and fermions were massless in the theory which is not true by experiment. This section describes spontaneous symmetry breaking which is the mechanism through which weak gauge bosons are believed to acquire their masses. To illustrate spontaneous symmetry breaking two examples are studied, a global gauge symmetry and a local gauge symmetry applied to $U(1)$ and $SU(2)$. First we start from a complex scalar field $\phi = (\phi_1 + i\phi_2)/\sqrt{2}$ described by the Lagrangian:

$$\mathcal{L} = (\partial_\mu \phi)^* (\partial^\mu \phi) - \mu^2 \phi^* \phi - \lambda(\phi^* \phi)^2. \tag{1.1}$$

where the first term is the kinetic energy term, the second term is the mass term and the third term is an interaction term. This Lagrangian is invariant under global gauge transformation $\phi \rightarrow e^{i\alpha}\phi$ with α independent on spatial coordinates. The Lagrangian is

$$\mathcal{L} = \frac{1}{2}(\partial_\mu \phi_1)^2 + \frac{1}{2}(\partial_\mu \phi_2)^2 - \frac{1}{2}\mu^2(\phi_1^2 + \phi_2^2) - \frac{1}{4}\lambda(\phi_1^2 + \phi_2^2)^2. \tag{1.2}$$

If the case where $\lambda > 0$ and $\mu^2 < 0$ is considered, the potential has a circle of minima at:

$$\phi_1^2 + \phi_2^2 = v^2 \text{ with } v^2 = -\frac{\mu^2}{\lambda}. \tag{1.3}$$

The fact that the minimum is not at zero is very interesting since for perturbation theory to work , the Lagrangian has to be defined around the global minimum that now is at v. Therefore the field needs to be translated to the new minimum by redefining it as:

$$\phi(x) = \sqrt{\frac{1}{2}}\,(v + \eta(x) + i\xi(x))\,. \tag{1.4}$$

Substituting the new field into the Lagrangian gives:

$$\mathcal{L}' = \frac{1}{2}(\partial_\mu \xi)^2 + \frac{1}{2}(\partial_\mu \eta)^2 + \mu^2 \eta^2 + \mathcal{O}(\eta^3, \xi^3, \eta^3, \xi^4). \tag{1.5}$$

The new Lagrangian has acquired a new mass term for the field η with mass $m_\eta = \sqrt{-2\mu^2}$ however there is no term like that for the field ξ. This is a consequence of the

Goldstone theorem that states that whenever a symmetry is spontaneously broken, always massless scalars occur in the new theory (the field ξ in this case).

The next step is to try to repeat the same process in a local gauge invariant $U(1)$ symmetry. So let us assume a scalar field interacting with an electromagnetic field. The theory is described by the Lagrangian:

$$\mathcal{L} = |D_\mu\phi|^2 - \mu^2\phi^*\phi - \lambda(\phi^*\phi)^2 - \frac{1}{4}F_{\mu\nu}F^{\mu\nu} \tag{1.6}$$

where $D_\mu = \partial_\mu - ieA_\mu$ is the covariant derivative and the transformation is $\phi \rightarrow e^{i\alpha(x)}\phi$ For $\mu^2 < 0$ the field has a new minimum υ exactly as before. Translating the Lagrangian to the new minimum gives:

$$\begin{aligned}\mathcal{L}' &= \frac{1}{2}(\partial_\mu\xi)^2 + \frac{1}{2}(\partial_\mu\eta)^2 + -\upsilon^2\lambda\eta^2 + \frac{1}{2}e^2\upsilon^2A_\mu A^\mu \\ &- e\upsilon A_\mu\partial^\mu\xi - \frac{1}{4}F_{\mu\nu}F^{\mu\nu} + \text{other terms}\end{aligned} \tag{1.7}$$

In this new Lagrangian, the field η has acquired a mass of $m_\eta = \sqrt{2\lambda\upsilon^2}$ but even more important is the fact that the field has also acquired a mass $m_A = e\upsilon$ that implies that the gauge boson of this theory became massive. The residual issue is that the field ξ is still massless that states that the Goldstone theorem could still stand. In addition the presence of a term $A_\mu\partial^\mu\xi$ is worrying since it introduces a coupling of the A field to the scalar field. In terms of physics, since the field acquired mass, it can have longitudinal polarization so one more degree of freedom was created. This cannot happen by translation of the variables so the particle spectrum in the Lagrangian is not correct. The additional term $A_\mu\partial^\mu\xi$ provides the term needed to make the polarization transverse. Therefore one of the fields in the Lagrangian does not correspond to a particle. The gauge invariance of the Lagrangian can be used to redefine it in a way that the particle content is more understandable. The complex field was defined as:

$$\phi = \sqrt{\frac{1}{2}}(\upsilon + \eta + i\xi) \approx \sqrt{\frac{1}{2}}(\upsilon + \eta)e^{\frac{i\xi}{\upsilon}} \tag{1.8}$$

That implies that the gauge can be picked differently so that

$$\phi \rightarrow \sqrt{\frac{1}{2}}(\upsilon + h(x))e^{i\theta(x)/\upsilon} \tag{1.9}$$

$$A_\mu \rightarrow A_\mu + \frac{1}{e\upsilon}\partial_\mu\theta \tag{1.10}$$

This particular choice of gauge transformation is known as the unitary gauge and is designed to make $h(x)$ real. Since θ appears as a phase factor, it will not appear in the final Lagrangian. Substituting into (1.6) gives for the final Lagrangian:

$$\begin{aligned}\mathcal{L}' &= \frac{1}{2}(\partial_\mu h)^2 - \lambda v^2 h^2 + \frac{1}{2}e^2 v^2 A_\mu^2 - \lambda v h^3 - \frac{1}{4}\lambda h^4 \\ &+ \frac{1}{2}e^2 A_\mu^2 h^2 + v e^2 A_\mu^2 h - \frac{1}{4}F_{\mu\nu}F^{\mu\nu}\end{aligned} \tag{1.11}$$

In this new Lagrangian the field has acquired a mass via the term $\frac{1}{2}e^2v^2A_\mu^2$. In addition, there is no massless Goldstone boson in the Lagrangian but a new scalar h with a mass of $m = \sqrt{2\lambda v^2}$ called the Higgs boson. So with this procedure the massless Goldstone boson has been converted to the additional longitudinal polarization degree of freedom so that the field acquires mass resulting in a Lagrangian with a scalar massive particle, a massive field and interaction terms. This procedure is known as the Higgs mechanism. For this mechanism to be the way vector bosons acquire mass in nature, a new particle must exist. The Higgs boson has not been yet discovered and it the search for it is the main purpose of this thesis.

The last step is to apply the Higgs mechanism in the electroweak model and create mass for the vector bosons and fermions. It is economical to use the same mechanism to give mass to the fermions but it could very well be that fermions acquire mass via a different mechanism. To apply the Higgs mechanism in the electroweak model we need to use the covariant derivative:

$$D_\mu = \partial_\mu + igT_a W_\mu^a + i\frac{g'}{2}B_\mu Y, \tag{1.12}$$

which corresponds to gauge invariance for the $SU(2)_L \times U(1)$ symmetry. Substituting (1.12) in the Lagrangian of the scalar field (Eq. (1.1)) gives:

$$\mathcal{L} = \left|\left(i\partial_\mu - gT_a W_\mu^a - g' B_\mu \frac{Y}{2}\right)\phi\right|^2 - V(\phi) \tag{1.13}$$

For this Lagrangian to be invariant, ϕ must be invariant under $SU(2) \times U(1)$ symmetry. A choice satisfying this is using an isospin doublet with Y$= 1$:

$$\phi = \begin{pmatrix}\phi^+ \\ \phi^0\end{pmatrix} \text{ with } \begin{matrix}\phi^+ = (\phi_1 + i\phi_2)/\sqrt{2} \\ \phi^0 = (\phi_3 + i\phi_4)/\sqrt{2}\end{matrix} \tag{1.14}$$

The new field will have a vacuum expectation value of

$$\phi_0 = \frac{1}{\sqrt{2}}\begin{pmatrix}0 \\ v\end{pmatrix} \tag{1.15}$$

Expanding the kinetic term of the scalar field and using the Pauli matrices as the generators of $SU(2)$ gives:

$$\left|\left(i\partial_\mu - gT_a W_\mu^a - \frac{g'}{2}B_\mu\right)\phi\right|^2$$

$$= \frac{1}{8}\left|\begin{pmatrix} gW_\mu^2 + g'B_\mu & g(W_\mu^1 - iW_\mu^2) \\ g(W_\mu^1 + iW_\mu^2) & -gW_\mu^3 + g'B_\mu \end{pmatrix}\begin{pmatrix} 0 \\ v \end{pmatrix}\right|^2$$

$$= \left(\frac{1}{2}vg\right)^2 W_\mu^+ W^{-\mu} + \frac{v^2}{8}\begin{pmatrix} g^2 & -gg' \\ -gg' & g'^2 \end{pmatrix}\begin{pmatrix} W^{3\mu} \\ B^\mu \end{pmatrix} \tag{1.16}$$

where we used the W raising and lowering operators defined as $W_\mu^\pm = (W_\mu^1 \pm iW_\mu^2)/\sqrt{2}$. The first term in this expression corresponds to a massive W boson with mass of $M_W = vg/2$. The matrix term can be expressed as a function of the fields A and Z by finding a transformation between W^3, B and A, Z that diagonalizes the matrix so that the elements in the diagonal are the mass terms $\frac{1}{2}M_Z^2 Z_\mu^2$, $\frac{1}{2}M_A^2 A_\mu^2$ for Z and A respectively. The transformation is:

$$A_\mu = \frac{g'W_\mu^3 + gB_\mu}{\sqrt{g^2 + g'^2}} \tag{1.17}$$

$$Z_\mu = \frac{gW_\mu^3 - g'B_\mu}{\sqrt{g^2 + g'^2}} \tag{1.18}$$

from which is derived that $M_A = 0$ (the photon remains massless!) and $M_Z = \frac{v}{2}\sqrt{g^2 + g'^2}$. Therefore all gauge bosons gained mass , the photon mass is zero and there is one additional scalar particle. The two couplings g and g' are related by electroweak unification via the weak mixing angle (θ_w) as $\tan\theta_w = g'/g$. The final step is to give mass to the fermions. The problem in $SU(2)_L \times U(1)$ was that the mass term $m\bar{\psi}\psi$ was not invariant. This can be solved by introducing a fermion coupling to the Higgs. In the case of the electron a term can be added in the Lagrangian of the form:

$$\mathcal{L}_e = -K\left[\left(\bar{\nu_e}\ \bar{e}\right)_L \begin{pmatrix} \phi^+ \\ \phi \end{pmatrix} e_R + \bar{e_R}\left(\phi^-\ \bar{\phi^0}\right)\begin{pmatrix} \nu_e \\ e \end{pmatrix}_L\right] \tag{1.19}$$

Substituting the form of the Higgs field in the unitary gauge the Lagrangian becomes

$$\mathcal{L}_e = -\frac{K}{\sqrt{2}}v\left(\bar{e_L}e_R + \bar{e_R}e_L\right) - \frac{K}{\sqrt{2}}\left(\bar{e_L}e_R + \bar{e_R}e_L\right)h \tag{1.20}$$

Picking the constant K such as that $m_e = Kv/\sqrt{2}$ gives a fermion mass term in the Lagrangian:

$$\mathcal{L}_e = -m_e \bar{e}e - \frac{m_e}{v}\bar{e}eh, \quad (1.21)$$

Therefore the electron has acquired mass and a new coupling has been introduced between the Higgs boson and the fermions proportional to the ratio of the electron and W mass, which is very small since m_e is small.

1.4 *Z* and Higgs Boson Production in Proton Collisions

The Standard Model describes all interactions between quarks and leptons. Those interactions result in production and decay of the $W^{\pm}$, Z and Higgs bosons. However to study scalar and vector boson production in proton collisions, the hard interaction between quarks and gluons has to be separated from the proton substructure. The quarks inside the proton are not free but are strongly bound, exchanging colored gluons. The continuous interactions between the quarks makes them virtual particles lying off their mass shell. Gluons are carrying about 50 % of the proton momentum and can produce additional quark–antiquark pairs (e.g. $s\bar{s}$). Therefore, the quark content of the proton is separated into the valence quarks (u, d), the sea quarks (u, d, s, c) and the gluons, together called partons.

Therefore, in a proton collision, any quark or gluon combination can contribute to the hard scattering process. The probabilities for a specific parton with a given momentum fraction of the proton to participate in the hard process are known as parton distribution functions and are measured using data from deep inelastic scattering experiments.

Therefore the center of mass energy $\sqrt{s}$ for each hard scattering process is related to the center of mass energy of the two colliding protons $\sqrt{S}$ as:

$$s = xyS \quad (1.22)$$

where x, y are the momentum fractions of the contributing partons. One important aspect of hadron colliders is that those parton distribution functions make them versatile machines for discovery although the hard scattering center of mass is not well defined. If the center of mass energy of the protons is high enough, depending on the parton momentum fractions, every collision will create particles of different masses and the possibility to observe new physics increases. This section describes the phenomenology of Z and Higgs bosons in proton collisions.

1.4.1 Z Boson Production in Proton Collisions

The Z boson is produced in proton collisions at leading order by fusion of quark–antiquark pairs (Fig. 1.2a) . This process can create a virtual photon that will decay to a fermion pair or a Z boson or a W^+W^- boson. Neglecting the virtual photon part and assuming massless fermions, the matrix element for the Z production is:

$$|\mathcal{M}|^2 = (s\sqrt{2}g_Z)^2 \frac{(g_V)^2 + (g_A)^2}{2} M_Z^2 = 32\frac{G_F}{\sqrt{2}}(g_V^2 + g_A^2)M_Z^4 \tag{1.23}$$

and the differential cross section at parton level is:

$$\sigma = 8\pi\frac{G_F}{\sqrt{2}}(g_V^2 + g_A^2)M_Z^2\delta(s - M_Z^2), \tag{1.24}$$

where s is the center of mass energy of the colliding quarks. What this formula implies is that to create a Z boson the two quarks must have a center of mass energy equal to the Z mass. In lepton colliders where elementary particles are colliding, the energy of the beams is set to the Z mass to create a Z boson. In hadron colliders the situation is different. In the LHC protons are colliding at a large constant energy of $\sqrt{S} = 7\,\text{TeV}$ in 2011. For the Z to be created, two partons from the proton interact. Those partons must be a quark–antiquark with suitable momentum fractions x, y such that $xyS = s$. Then the parton distribution functions are integrated for x, y to give the full cross section:

$$\sigma = \frac{8\pi}{3}\frac{G_F}{\sqrt{2}}\int \mathrm{d}x\mathrm{d}y \sum (g_V^2 + g_A^2)xySf_q(x)f_{\bar{q}}(y) \tag{1.25}$$

where the integration is performed under the constraint $xyS = M_Z^2$ and f_q are the parton distribution functions. The sum runs over all quarks.

The Drell–Yan process of Fig. 1.2a is the simplest diagram that can create a Z boson. In this process, the Z has no transverse momentum but is created and moving along the beamline. There is a significant contribution from higher order diagrams

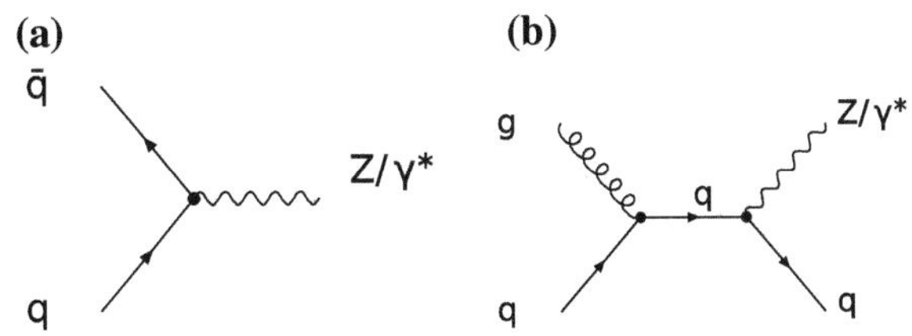

Fig. 1.2 Z boson production mechanisms in hadron colliders. **a** $Z + 0$ jet production; **b** $Z + 1$ jet production

such as those shown in Fig. 1.2b resulting in an additional parton in the final state that hadronizes to a jet . Those types of events are also observed and constitute very important background in Higgs searches as it is described in later chapters. In terms of the Z decay signature, the Z boson decays primarily to hadrons or lepton pairs according to the V–A couplings. Z decays to a pair of leptons about 3 % of the time and the most straightforward final state for Z measurement is a muon or electron pair. In this thesis, a measurement of the Z cross section is presented via a Z decays to tau pairs.

1.4.2 Higgs Boson Production in Proton Collisions

The Higgs boson couples to mass, therefore, it has large couplings to W and Z boson pairs. In terms of the fermions, the introduced Yukawa couplings impose a dependence of the coupling as m_f^2, therefore the Higgs coupling is enhanced for heavy quarks (especially t) and leptons (especially τ). The dominant production of Higgs bosons in proton collisions at high energies is via gluon fusion (Fig. 1.3a)

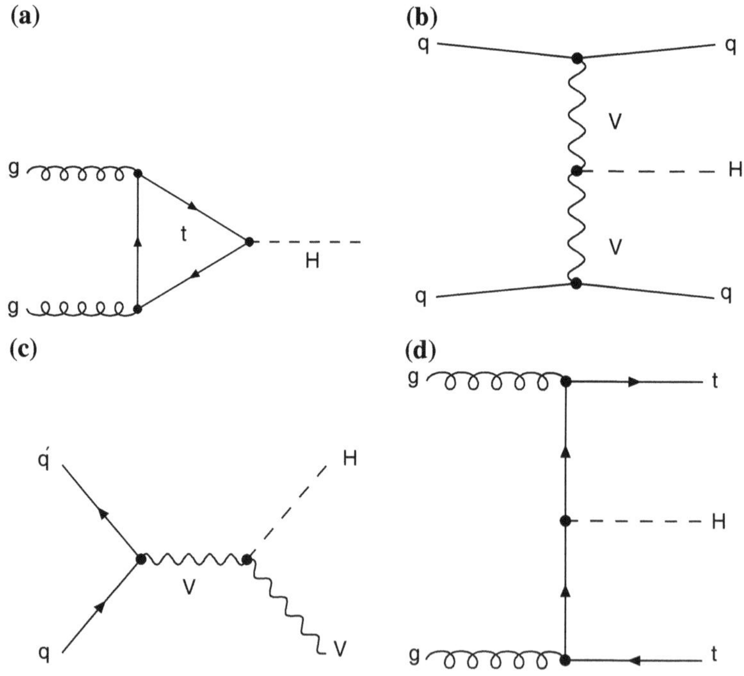

Fig. 1.3 Standard model Higgs production mechanisms in proton collisions. **a** Gluon fusion. **b** Vector boson fusion. **c** Associated production with vector boson. **d** Associated production with top quark pair

through a fermion loop. Since the Higgs boson couples to the fermion in the loop, the loop contribution is dominated by the top quark and the high parton luminosity of for gluons ensures significant cross section. The second most dominant mechanism is via vector boson fusion (Fig. 1.3b) where Higgs is produced via the fusion of a pair of W or Z bosons. Those bosons are initially radiated by a pair of quarks. This signature is extremely important in this analysis because it introduces a final state with two forward jets from the outgoing quarks. Those jets can be detected and enhance the sensitivity. The other two main production mechanisms involve associated production of Higgs with a vector boson (Fig. 1.3c) and associated production of Higgs with a top quark pair (Fig. 1.3d). The production cross section and the decay branching fractions as a function of the Higgs mass are presented in Fig. 1.4. In terms of decays, Higgs decays primarily to fermion or boson pairs and the branching ratio is enhanced for heavier particles. For low Higgs mass, the primary decays are via $b\bar{b}$ and $\tau\tau$ while for higher masses the WW and ZZ final states dominate. For $m_H \approx 2m_t$ the Higgs decay to top pairs also contributes. Despite the structure of the couplings, the most sensitive modes for the Higgs depend on experimental reasons. For example, at low mass, the $b\bar{b}$ final state can only contribute via associated production (WH, ZH) since the $pp \to b\bar{b}$ background is overwhelming. In the same sense, $H \to \tau\tau$

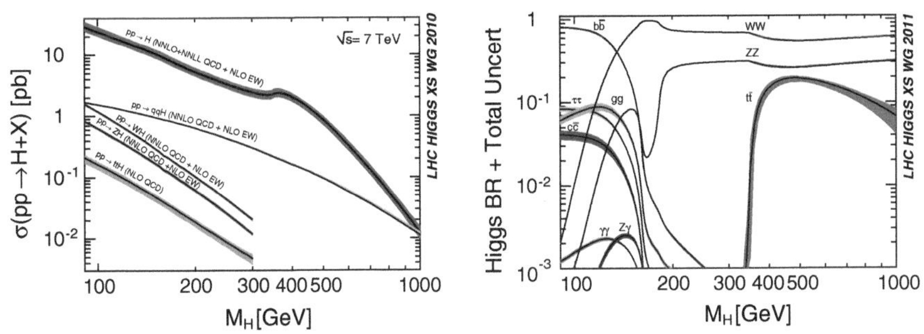

Fig. 1.4 Higgs cross section for the different production mechanisms (*left*) and branching fractions (*right*) as a function of the Higgs boson mass [15]

Table 1.1 Resonances and branching ratios of the dominant hadronic decays of the τ-lepton [17]

Decay mode	Resonance	Mass (MeV/c^2)	Branching ratio (%)
$\tau^- \to h^- \nu_\tau$			11.6
$\tau^- \to h^- \pi^0 \nu_\tau$	ρ	770	26.0
$\tau^- \to h^- \pi^0 \pi^0 \nu_\tau$	$a1$	1200	10.8
$\tau^- \to h^- h^+ h^- \nu_\tau$	$a1$	1200	9.8
$\tau^- \to h^- h^+ h^- \pi^0 \nu_\tau$			4.8
Total			63.1
Other hadronic modes			1.7

The decay products listed correspond to a negatively charged τ-lepton; the table is identical under charge conjugation dominantly contributes via VBF due to the overwhelming $Z \to \tau\tau$ background. On the other hand ZZ and WW final states also contribute to lower mass searches because of the very clean signatures $ZZ \to 4\ell$ and $WW \to 2\ell 2\nu$. Finally $H \to \gamma\gamma$ has very low branching ratio but the relatively clean two photon signature provides the best sensitivity at low mass.

1.5 The Tau Lepton

The τ lepton is the heaviest of the leptons. It was discovered by Perl et al. in Mark I experiment [16] at Stanford Linear Accelerator Center. The observation was made using events with one muon and electron in the final state. Due to its high mass of 1.779 GeV it can decay to many lighter particles. Taus decay via the weak interactions emitting a virtual W boson and a tau neutrino. The virtual W boson can hadronize into some mesons or decay semileptonically to muons or electrons. About 65 % of the time tau decays hadronically while the other cases involve muons and electrons. The hadronic decays of the tau lepton are usually via the $\rho(770)$ resonance where $\rho \to \pi^+\pi^0$ or via the $a_1(1200)$ resonance where $a_1 \to \pi^+\pi^-\pi^+$ or $a_1 \to \pi^+\pi^0\pi^0$. The tau decays and branching ratios are presented in Table 1.1. One characteristic of taus produced from heavy particle decays such as the Z or Higgs bosons is that the decay products are boosted. Therefore the tau appears in the detector as a highly collimated collection of particles and this signature is used to detect them. The identification of hadronic decays of taus is crucial to identify scalar and vector bosons decaying to tau pairs and in this thesis a novel method is proposed for this purpose.

References

1. C.N. Yang, *Elementary Particles* (Princeton University Press, Princeton, 1961)
2. D. Griffiths, *Introduction to Elementary Particles* (Wiley, New York, 1987)
3. S.L. Glashow, J. Iliopoulos, L. Maiani, Weak interactions with Lepton–Hadron symmetry. Phys. Rev. D **2**, 1285–1292 (1970)
4. S.L. Glashow, Partial symmetries of weak interactions. Nucl. Phys. **22**, 579–588 (1961)
5. S. Weinberg, A model of leptons. Phys. Rev. Lett. **19**, 1264–1266 (1967)
6. N. Svartholm, A. Salam, Elementary particle physics: relativistic groups and analycity, in *8th Nobel Symposium*, Stockholm, Almquvist and Wiksell, 1968, p. 367
7. F. Englert, R. Brout, Broken symmetry and the mass of gauge vector mesons. Phys. Rev. Lett. **13**, 321–323 (1964)
8. G. Guralnik, C. Hagen, T. Kibble, Global conservation laws and massless particles. Phys. Rev. Lett. **13**, 585 (1964)
9. P.W. Higgs, Broken symmetries, massless particles and gauge fields. Phys. Lett. **12**, 132 (1964)
10. P.W. Higgs, Broken symmetries and the masses of gauge bosons. Phys. Rev. Lett. **13**, 508 (1964)
11. P.W. Higgs, Spontaneous symmetry breakdown without massless bosons. Phys. Rev. **145**, 1156 (1966)

12. M.H. Ahn et al., Measurement of neutrino oscillation by the K2K experiment. Phys. Rev. D **74**, 072003 (2006)
13. P. Adamson et al., A study of muon neutrino disappearance using the Fermilab main injector neutrino beam. Phys. Rev. D **77**, 072002 (2008)
14. P. Adamson et al., Measurement of neutrino oscillations with the MINOS detectors in the NuMI beam. Phys. Rev. Lett. **101**, 131802 (2008)
15. LHC Higgs Cross Section Working Group, S. Dittmaier, C. Mariotti, G. Passarino, R. Tanaka et al. (eds.), Handbook of LHC Higgs Cross Sections: Inclusive Observables (2011)
16. M.L. Perl, G.S. Abrams, A. Boyarski, M. Breidenbach et al., Evidence for anomalous lepton production in e+–e− annihilation. Phys. Rev. Lett. **35**, 1489–1492 (1975)
17. K. Nakamura, Particle Data Group, Review of particle physics. J. Phys. G Nucl. Part. Phys. **37**(7A), 075021 (2010)

Chapter 2
Supersymmetry and the MSSM

The Standard Model is a very successful theory in predicting the experimental observations through the latest years. With a possible discovery of the Higgs boson, the SM will be a theory of almost everything. However questions arise when the Standard Model is seen as part of a larger unified theory. The masses of the gauge bosons in SM are derived by the vacuum expectation value of the Higgs boson which it was found in the previous chapter to be

$$v = 246\,\text{GeV} \tag{2.1}$$

The mass of the W/Z and Higgs depends on this value. A first question is what is happening if loops are included in the diagrams. One example is the Higgs self interaction diagram in Fig. 2.1 on the left.

This diagram will correct the Higgs mass in higher orders by a factor of

$$\lambda \int^{\Lambda} \mathrm{d}^4 k \frac{1}{k^2 - m_h^2} \tag{2.2}$$

which seems to be quadratically divergent. However, the SM is a renormalizable theory. This means that the Lagrangian term $-\mu^2$ can be picked so that it depends on the cut-off Λ and if $\Lambda \to \infty$, the renormalized coefficient has the desired value. The problem, however, arises when we see the SM as a part of a larger theory that appears in a scale Λ where the SM needs to be modified. One example is what happens at the Planck scale where quantum gravity becomes important. If the integral in Eq. 2.2 stops at a finite scale Λ the correction term to the mass is positive and proportional to $\lambda\Lambda^2\phi^\dagger\phi$. Therefore if there is new physics at a higher scale the SM is going to collapse. This problem is known as the hierarchy problem [1]. One important step in solving the hierarchy problem is to consider the one loop correction to the Higgs mass but via a fermion loop this time (Fig. 2.1 on the right). This term is:

M. Bachtis, *Heavy Neutral Particle Decays to Tau Pairs*,
Springer Theses, DOI: 10.1007/978-3-319-03257-3_2,

Fig. 2.1 Scalar (*left*) and fermion (*right*) loop corrections to the Higgs mass

$$\left[-g^2 \int^{\Lambda} \frac{d^4k}{\not{k}\ \not{k}}\right] \phi^\dagger \phi \approx = -g^2 \Lambda^2 \phi^\dagger \phi \tag{2.3}$$

Combining this term with the one derived from the scalar loop gives:

$$(\lambda - g^2)\Lambda^2 \phi^\dagger \phi \tag{2.4}$$

So if $g^2 = \lambda$ the quadratic divergence disappears. Therefore if for each scalar there was a fermion partner with $g^2 = \lambda$ the divergences would not occur any more. The existence of such a particle partner imposes a new symmetry between fermions and bosons known as Supersymmetry [2]. To define Supersymmetry (SUSY) we need a set of generators for the symmetry that can turn a bosonic state to a fermionic state and vice versa. That implies that the generators themselves must carry spin of 1/2. The simplest selection of SUSY generators is a 2-component Weyl spinor Q and its conjugate $\bar{Q}$ such that:

$$Q|\text{boson}>= |\text{fermion}> \quad Q|\text{fermion}>= |\text{boson}> \tag{2.5}$$

The algebra of the generators can be written in terms of the anti-commutators as:

$$Q_\alpha, Q_\beta = \bar{Q}_{\dot{\alpha}}, \bar{Q}_{\dot{\beta}} = 0 \tag{2.6}$$

$$[Q_\alpha, P_\mu] = 0 \tag{2.7}$$

$$Q_\alpha, \bar{Q\beta} = 2\sigma^\mu_{\alpha\beta} P_\mu \tag{2.8}$$

where P_μ is the translation generator (momentum) and α, β take values of 1,2. From the commutation relations it can be shown that:

$$P^\mu P_\mu Q|b>= P^\mu P_\mu|f>= m_{\tilde{f}}^2|f> \tag{2.9}$$

but also

$$P^\mu P_\mu Q|b>= QP^\mu P_\mu|b>= m_{\tilde{b}}^2 Q|b>= m_{\tilde{b}}^2|f> \tag{2.10}$$

Therefore the masses of the boson and the fermions must be the same. This has not been observed experimentally which leads to the conclusion that if SUSY exists it must be broken so that the masses of the bosons and the corresponding fermions must be different.

2.1 The Minimal Supersymmetric Standard Model

The Minimal Supersymmetric Standard Model (MSSM) is a straightforward supersymmetric extension of the SM. It is designed to satisfy the phenomenological constraints imposed by the experimental observations. SUSY transformations do not act in the $SU(3)$, $SU(2)$, $U(1)$ degrees of freedom. For example a lepton doublet cannot have the Higgs boson as a super-partner since the Higgs doublet doesn't carry lepton number which needs to be conserved. Therefore, the spectrum of SM particles must be doubled in the MSSM. There are two types of super-multiplets in SUSY theories: Chiral super-multipltes that contain one fermion and a complex scalar boson and gauge supermultiplets that contain one vector boson and one fermion. All fermions in the SM are parts of chiral super-multiplets. Their scalar super-partners are called squarks and sleptons and are coupling to the electroweak interactions exactly as the nominal SM particles. The SM vector boson fields are members of gauge super-multiplets and their fermionic super-partners are known as gauginos (gluinos, winos, binos). Photinos and zinos are defined as linear combinations of winos and binos, exactly as in the SM case. For the Higgs sector two Higgs doublets are needed and their corresponding fermion super-partners (Higgsinos). The Higgs sector is described in detail in the next section.

In SUSY, particles are assigned a quantum number known as R-parity defined as:

$$R = (-1)^{3B+L+2S}, \tag{2.11}$$

where B is the baryon number, L is the lepton number and S is the spin. R is $+1$ for usual SM particles and -1 for SUSY particles. The R parity is introduced to forbid couplings in the theory that allow the baryon and lepton numbers to not be conserved. Such couplings would allow decay of the proton. Conservation of R parity introduces additional phenomenological constraints such as the fact that the lighter supersymmetric particle (LSP) is stable. This is interesting in the sense that LSP could be a very good candidate for dark matter. In addition, the production of supersymmetric particles in colliders must conserve the R parity therefore they are always produced in pairs.

2.2 The Higgs Sector in the MSSM

In the MSSM, the Higgs sector is described by a model consisting of two Higgs doublets.

$$\Phi_1 = \begin{pmatrix} \phi_1^{0*} \\ -\phi_1^- \end{pmatrix} \quad \Phi_2 = \begin{pmatrix} \phi_2^{\dagger} \\ \phi_2^0 \end{pmatrix} \tag{2.12}$$

The Φ_1 doublet has hypercharge of -1 and gives masses to the down-type quarks and charged leptons while the Φ_2 doublet gives masses to the up-type quarks. In the

MSSM two doublets are needed two cancel out anomalies in triangle diagrams with three bosons and a triangle loop. Those diagrams cancel in the SM due to the equal number of quark and lepton generations. In the same manner, two Higgs doublets are needed to cancel out the anomalies in the triangle loops with higgsinos in the loop. Each Higgs doublet has a different vacuum expectation value. Translating the fields at their minima gives the following mass eigenstates:

$$\begin{pmatrix} H \\ h \end{pmatrix} = \sqrt{2}\begin{pmatrix} \cos\alpha & \sin\alpha \\ -\sin\alpha & \cos\alpha \end{pmatrix}\begin{pmatrix} \mathrm{Re}\phi_1^{0*} - v_1 \\ \mathrm{Re}\phi_2^0 - v_2 \end{pmatrix} \tag{2.13}$$

$$A = \sqrt{2}(\sin\beta \mathrm{Im}\phi_1^{0*} + \cos\beta \mathrm{Im}\phi_2^0), \tag{2.14}$$

$$H^- = (H^+)^* = -\phi_1^- \sin\beta + \phi_2^- \cos\beta \tag{2.15}$$

Therefore in the MSSM the Higgs sector consists of two scalar bosons h, H, a pseudo-scalar A and a pair of charged Higgs bosons $H^{\pm}$. The angle β is related to the vacuum expectation values of the two Higgs doublets as

$$\tan\beta = \frac{v_1}{v_2} \tag{2.16}$$

At tree level, the MSSM Higgs sector is described only by two parameters which can be the mass of the pseudo-scalar A and the $\tan\beta$. The masses of the other neutral bosons are given by:

$$m_h^2 = \frac{1}{2}(m_A^2 + m_Z^2 - \sqrt{(m_A^2 - m_Z^2)^2 + 4m_Z^2 m_A^2 \sin^2(2\beta)} \tag{2.17}$$

$$m_H^2 = \frac{1}{2}(m_A^2 + m_Z^2 + \sqrt{(m_A^2 - m_Z^2)^2 + 4m_Z^2 m_A^2 \sin^2(2\beta)} \tag{2.18}$$

while the mass of charged Higgs in tree level is given by

$$m_{H^{\pm}}^2 = m_A^2 + M_W^2 \tag{2.19}$$

To make reliable phenomenological predictions, loop corrections have to be included that depend on the masses of the SUSY particles. Therefore specific benchmark scenarios are chosen for the MSSM based on the SUSY parameters. This thesis uses the $m_h^{\max}$ scenario [3] which is defined by:

$$\begin{aligned} M_{\mathrm{SUSY}} &= 1\,\mathrm{TeV},\ X_{\mathrm{t}} = 2M_{\mathrm{SUSY}},\ \mu = 200\mathrm{GeV}, \\ M_{\tilde{g}} &= 800\,\mathrm{GeV},\ M_2 = 200\,\mathrm{GeV},\ A_{\mathrm{b}} = A_{\mathrm{t}}, \end{aligned} \tag{2.20}$$

where M_{SUSY} denotes the common soft-SUSY-breaking squark mass of the third generation, $X_t = A_t - \mu/\tan\beta$ the stop mixing parameter, A_t and A_b the stop and sbottom trilinear couplings, respectively, μ the Higgsino mass parameter, $M_{\tilde{g}}$ the gluino mass, and M_2 the SU(2)-gaugino mass parameter. M_1 is fixed via the GUT-relation $M_1 = 5/3M_2 \sin\theta_w/\cos\theta_w$.

2.3 MSSM Higgs Production in Proton Collisions

The two doublet structure of the Higgs sector in the MSSM is responsible for some very interesting phenomenological effects that do not appear in the SM. The dominant production mechanisms are gluon fusion and associated production of b quarks (Fig. 2.2)). For large values of $\tan\beta$, $\mathcal{O}(10)$, the couplings to the down-type fermions is enhanced practically for the heaviest ones (b, τ) resulting in higher branching ratios for the $\phi \to \tau^+\tau^-$ and $\phi \to b\bar{b}$ final states that are almost constant as a function of m_A at the values of 10 % and 90 % respectively.

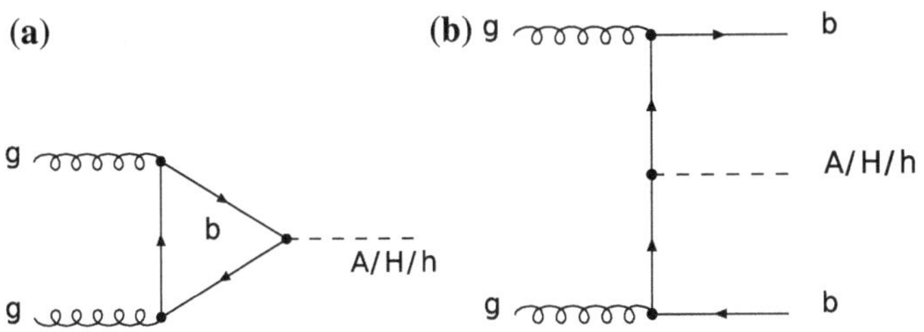

Fig. 2.2 Dominant MSSM Higgs production mechanisms in proton collisions. **a** Gloun fusion **b** Associated production with b quarks

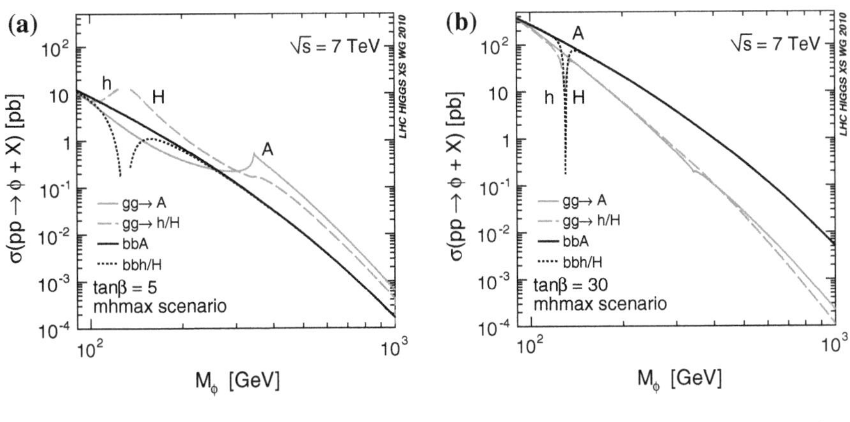

Fig. 2.3 Production cross section for different MSSM Higgs bosons and production mechanisms [4]. **a** $\tan\beta = 5$ **b** $\tan\beta = 30$

In addition, the top loop in the gluon fusion is augmented by an enhanced bottom loop resulting in enhancement of the gluon fusion cross section by a factor proportional to tan β^2. Moreover, due to the increased b couplings, the associated production with b quarks becomes a dominant mechanism. The presence of b quarks in the final state makes possible to identify them and enhance sensitivity against the $Z \to \tau\tau$ background. Figure 2.3 shows the production cross section for two different values of tan β. The enhancement of the cross section at higher tan β, the high branching ratio to di-tau final state and the presence of the associated production with b quarks makes the MSSM Higgs search extremely promising at the LHC.

References

1. E. Witten, Mass hierarchies in supersymmetric theories. Phys. Lett. B **105**, 267 (1981)
2. S.P. Martin, A supersymmetry primer. (1997)
3. M.S. Carena et al., MSSM Higgs boson searches at the Tevatron and the LHC: impact of different benchmark scenarios. Eur. Phys. J. **C45**, 797–814 (2006)
4. LHC Higgs Cross Section Working Group, in *Handbook of LHC Higgs Cross Sections: Inclusive Observables*, ed. by S. Dittmaier, C. Mariotti, G. Passarino, R. Tanaka et al. (CERN, Geneva, Switzerland, 2011)

Chapter 3
Experimental Setup

3.1 The Large Hadron Collider

The Large Hadron Collider(LHC) is a superconducting hadron accelerator and collider consisting of two rings that are installed in the existing 26.7 km tunnel that was constructed between 1984 and 1989 for the LEP machine [1]. Unlike LEP which collides electrons and positrons, LHC is designed as a particle-particle collider, consisting of two rings of counter rotating beams of particles. The tunnel (Fig. 3.1) has a diameter of 3.7 m, housing a twin-bore magnet which provides both rings in the same structure. The other main characteristic of the LHC is that it utilizes the latest technology on superconducting magnets that operate at the 2 K regime.

3.1.1 Performance Goals and Constraints

The LHC is designed to reach a center of mass energy up to 14 TeV. For any physics process the number of events generated by LHC collisions is given by:

$$N_{event} = \mathcal{L}\sigma_{event}, \tag{3.1}$$

where $\mathcal{L}$ is the instantaneous machine luminosity and σ_{event} is the production cross section for this specific physics process. The machine luminosity depends on the beam parameters and for a Gaussian distributed beam is given by:

$$\mathcal{L} = F\frac{N_b^2 n_b f_{rev}\gamma_r}{4\pi\epsilon_n\beta*}, \tag{3.2}$$

where F the geometric luminosity reduction factor due to the crossing angle at the interaction point (IP), N_b is the number of particles per bunch, n_b the number of bunches per beam, f_{rev} the revolution frequency, γ_r the relativistic gamma factor,

M. Bachtis, *Heavy Neutral Particle Decays to Tau Pairs*,
Springer Theses, DOI: 10.1007/978-3-319-03257-3_3,

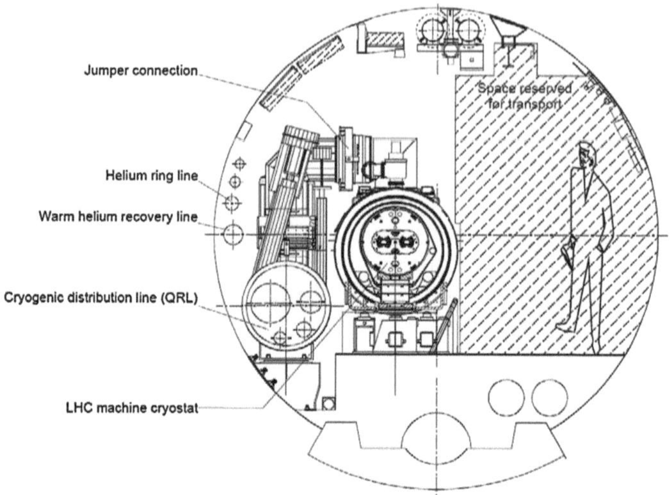

Fig. 3.1 Cross section of the LHC tunnel

ϵ_n the normalized transverse beam emitance and $\beta*$ the beta function at the collision point The beam emitance and the $\beta*$ are determined from the single particle transverse motion described by the equation:

$$x(s) = A\sqrt{\beta(s)}\cos\left[\psi(s) + \delta\right], \tag{3.3}$$

where s is the path along beam direction and A, δ are constants of integration defined by boundary conditions. The amplitude function, β, describes the amplitude of the motion in the beam line. The phase depends on β and advances as $d\psi/ds = 1/\beta$. The amplitude function at the IPs where magnets are configured to focus the beams is designated as $\beta*$. The number of transverse oscillations per rotation is denoted as the tune (ν). The phase space of the motion in the transverse plane is described by an ellipse in the (x, x') plane where $x' = dx/ds$. This ellipse has a total area of πA^2. If we consider a particle ensemble populating the phase space we define as emitance the area populated by this particle ensemble. This area depends only on the beam energy. For a Gaussian one dimensional beam, the emitance is defined as:

$$\epsilon = \pi\frac{\sigma^2}{\beta} \tag{3.4}$$

The geometric luminosity reduction factor is given by:

$$F = \left(1 + \left(\frac{\theta_c\sigma_z}{2\sigma*}\right)^2\right)^{-1/2}, \tag{3.5}$$

where θ_c is the full crossing angle in the IP, σ_z is the RMS bunch length and $\sigma*$ the transverse RMS beam size at the IP. The LHC is designed to operate at a maximum luminosity of the order of $\mathcal{L} = 10^{34}\text{cm}^{-2}\text{s}^{-1}$. That implies nominal operation with large number of bunches (2,808 per beam) and bunch crossings spaced by 25 ns. To reach this luminosity, very large proton density is required. Therefore, proton-proton beams are used providing the highest beam currents.

The limitations to the maximum instantaneous luminosity come from several factors. The particle density per bunch is mainly limited by the non-linear beam-beam interactions between particles when beams collide with each other. Beam-beam interactions result in modification of the tune of the particles. which is expressed via the linear tune shift given by:

$$\xi = \frac{N_b r_p}{4\pi\epsilon_n}. \tag{3.6}$$

In this equation, r_p is the classical radius of the proton ($r_p = e^2/4\pi\epsilon_0 m_p c^2$). The linear tune shift should not exceed 0.015 when summed over all IPs in the machine. Another parameter that affects luminosity is the mechanical aperture which is given by the beam screen dimensions. The beam screen in the LHC has a width of 2×22 mm and a height of 2×17.3 mm. Setting the aperture at 10σ in terms of the RMS beam size and accounting for imperfections in design and alignment results in a maximum nominal beam size of 1.2 mm. Combined with a maximum β-function of 180 m in the arcs, this implies a maximum emitance of $\epsilon_n = 3.75\,\mu\text{m}$. This parameter in combination with the linear beam-beam tune shift, limits the particle density to $N_b = 1.15 \times 10^{11}$. Furthermore, the mechanical aperture limits the $\beta*$ value at the IPs resulting in lower peak luminosity. Other design parameters that affect the peak luminosity are related to the maximum dipole field, the maximum energy that can be stored in the machine, the heat load, the quality of the magnetic field and the beam instabilities due to electromagnetic interactions between particles and between particles and the conducting boundaries of the vacuum. The dipole field corresponding to a beam energy of 7 TeV is 8.33 T. However, the operating dipole field depends on the heat load and temperature margins inside the cryo-magnets and therefore on beam losses, causing high dipole field to require very low beam losses during operation. Heat load refers to heat deposition that is absorbed by the cryogenic system and it is usually due to energy losses and synchrotron radiation. Synchrotron radiation is given by Larmor's formula:

$$P = \frac{1}{6\pi\epsilon_0}\frac{e^2\alpha^2}{c^3}\gamma^4, \tag{3.7}$$

where α is the centripetal acceleration. Although the LHC is a proton collider, 7 kW have to be removed at 1.8 K around the ring due to heating from synchrotron radiation which is an experimental challenge for the LHC.

Other parameters that limit the peak luminosity are related to the operation of the machine. The most important metric for LHC Physics program is the integrated luminosity for a run which is defined as:

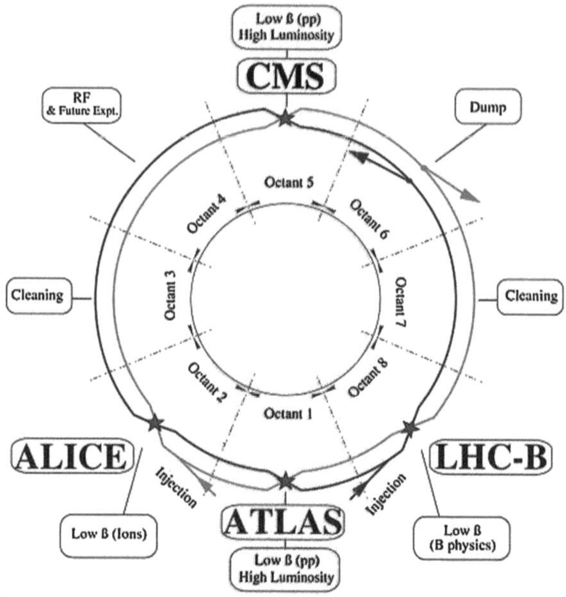

Fig. 3.2 Schematic of the layout of the LHC. Beam one is rotating clockwise while beam two is rotating anticlockwise

$$\mathcal{L}_{int} = L_0 \tau_L \left[1 - e^{-T_{run}/\tau_L}\right], \tag{3.8}$$

where L_0 is the initial instantaneous luminosity, τ_L is the net luminosity lifetime and T_{run} is the total length of the luminosity run. The luminosity lifetime depends on collisions that can degrade the beam and other effects like beam-beam interactions, radiation and scattering of particles in the residual gas. The average design luminosity lifetime for LHC is

$$\tau_L = 14.9\,\text{h} \tag{3.9}$$

The LHC is designed for long term operation. It will operate for about 200 days per year. One significant parameter for high integrated luminosity is the turn-around time, which is the time from the end of the run to the beginning of a new one. Ideally this time is about 1.15 h due to magnet hysterysis, however experience in the operation of LHC and other accelerators has shown that a more realistic value is of the order of 7 h. All those parameters limit the maximum integrated luminosity that the LHC provides in design conditions to 80–120 fb^{-1} per year.

3.1.2 Layout and Operation

The LHC Layout is depicted in Fig. 3.2. The LHC consists of eight arcs and eight straight sections. Each straight section is 528 m long and can have a collision point

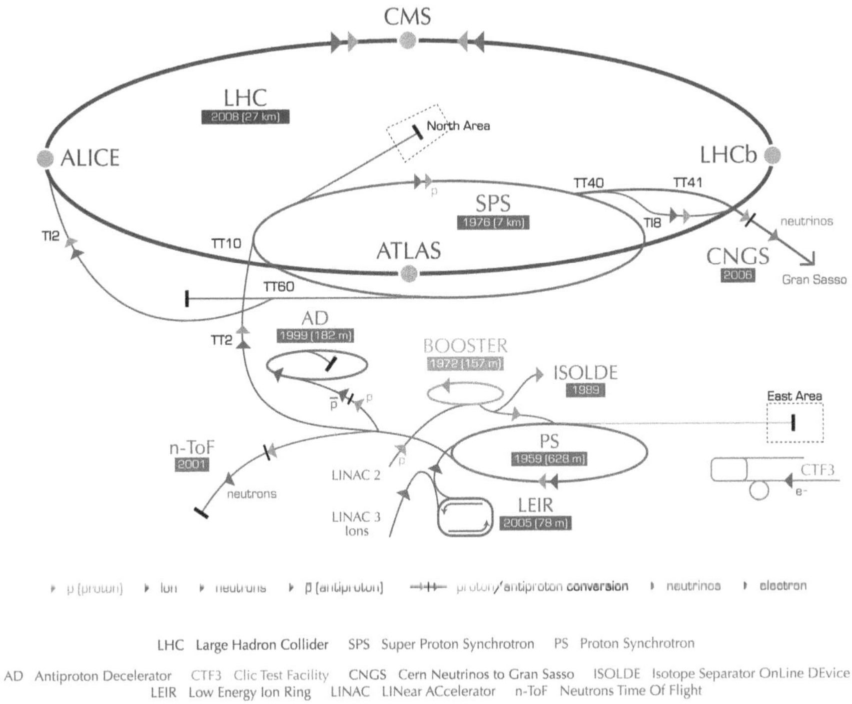

Fig. 3.3 The CERN accelerator complex

that serves to host either detectors or specific LHC equipment. The main experimental access points are Points 1, 2, 5 and 8. Points 1 and 5 host the high luminosity experiments ATLAS and CMS respectively. Points 2 and 8 host the injection systems for each beam. Points 3 and 7 host two collimation systems and Point 4 contains two RF systems. The straight section at Point 6 is used for the beam dump system. The LHC contains 23 arc cells which are 106.9 m long and consist of two 53.45 m long half cells, each of which contains one 5.355 m long cold mass, a short straight station(SSS) assembly and three 14.3 m long dipole magnets. The LHC arc cell layout is optimized for a maximum integrated dipole field along the arc with a minimum number of magnet interconnections and smallest possible beam envelopes.

Particles cross a large accelerator complex before reaching the final LHC energies (Fig. 3.3). Protons are accumulated using hydrogen atoms after stripping off electrons. Protons are first accelerated in LINAC2, a linear accelerator that consists of RF cavities up to an energy of 50 MeV. They are then injected to the Proton Synchrotron Booster(PSB) which ramps up the energy to 1.4 GeV, also by using RF cavities. In the PSB beams start to get squeezed and each of the six bunches of protons in the PSB is split into three. The next stage of accelerator is the Proton Synchrotron (PS). In the PS the energy is increased to 24 GeV and each bunch is split into two bunches twice. In addition, using an 80 MHz RF system, bunches are shortened so they can

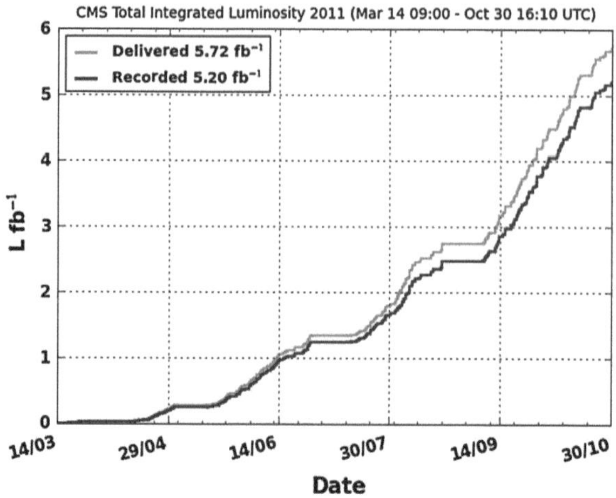

Fig. 3.4 LHC Delivered and CMS recorded luminosity in 2011 proton physics run

fit into the 200 MHz brackets of the next RF stage. The final stage before injection into the LHC is the Super Proton Synchrotron (SPS). The SPS increases the energy up to 450 GeV and then the beams enter the LHC ring.

3.1.3 Operating Conditions in 2011 Run

During 2011 run, LHC ramped up from very low luminosity to high luminosities of $\mathcal{L} = 2 \times 10^{33}\text{cm}^{-2}\text{s}^{-1}$. The increase in luminosity was performed by increasing the number of bunches but also optimizing optics to minimize $\beta*$. Figure 3.4 shows the integrated luminosity of the LHC during the 2011 Physics run.

One very important parameter for the experiments is the number of pile-up (PU) interactions; the number of interactions in a bunch during collision. Even if LHC experiments are designed to handle large multiplicities, large number of interactions complicates data analysis. The total event rate, as discussed in Sect. 3.1.1 is given by

$$R = \mathcal{L}\sigma_T, \tag{3.10}$$

where σ_T is the total pp inelastic cross section. Then the mean rate of events from collisions of a bunch is given by

$$N = \frac{R}{R_B}, \tag{3.11}$$

where R_B is the LHC crossing rate. Only a fraction of all LHC bunches are full (f_B), therefore the mean number of interactions per crossing is defined as

$$\mu = \frac{\mathcal{L}\sigma_T}{R_B f_B} \tag{3.12}$$

In the highest luminosity achieved by the LHC in the 2011 run thus far, the average number of interactions per crossing reached the value of $\mu = 16.5$ which makes trigger, particle identification and data analysis very challenging.

3.2 The Compact Muon Solenoid Experiment

The Compact Muon Solenoid (CMS) detector [2] is a general purpose experiment installed in Point 5 of the LHC ring, near the village of Cessy in France. CMS is designed to exploit the full physics reach of the LHC. Therefore it will operate in very high luminosity conditions. As described in Sect. 3.1, LHC will be delivering about 20 interactions per crossing in the center of the detector and to separate those interactions, excellent spatial resolution is needed close to the interaction point. In addition, due to the very high crossing rate ($R_B = 40$ MHz at design), very high speed of the detector electronics is required for time separation of the different crossings colliding every 25 ns at LHC design. Therefore, the general concept of a detector to operate in LHC conditions requires a large set of channels capable of operating in close synchronization and high speed.

CMS fulfills the following requirements:

- Good charged particle momentum resolution and reconstruction efficiency in the inner tracker, requiring high granularity for efficient reconstruction of a large number of charged particles near each other, especially near the interaction point.
- Good electromagnetic energy resolution and wide angular coverage for reconstructing electrons and photons.
- Good hadronic calorimeter energy resolution and hermetic coverage for providing good missing transverse energy performance and multi-jet mass resolution.
- Efficient identification of muons with good resolution over broad momenta ranges and angles.

To achieve the above goals CMS uses a large superconducting solenoid magnet that envelopes a large silicon tracker and both electromagnetic (ECAL) and hadronic (HCAL) calorimeters. Outside the solenoid there is an iron yoke for control of the magnetic flux return with multiple layers of muon detectors installed in it. A schematic view of the CMS detector is presented in Fig. 3.5.

3.2.1 Coordinate System

CMS uses the coordinate system with respect to the nominal collision point in the center of the detector. The x-axis points to the center of the LHC ring, the y-axis

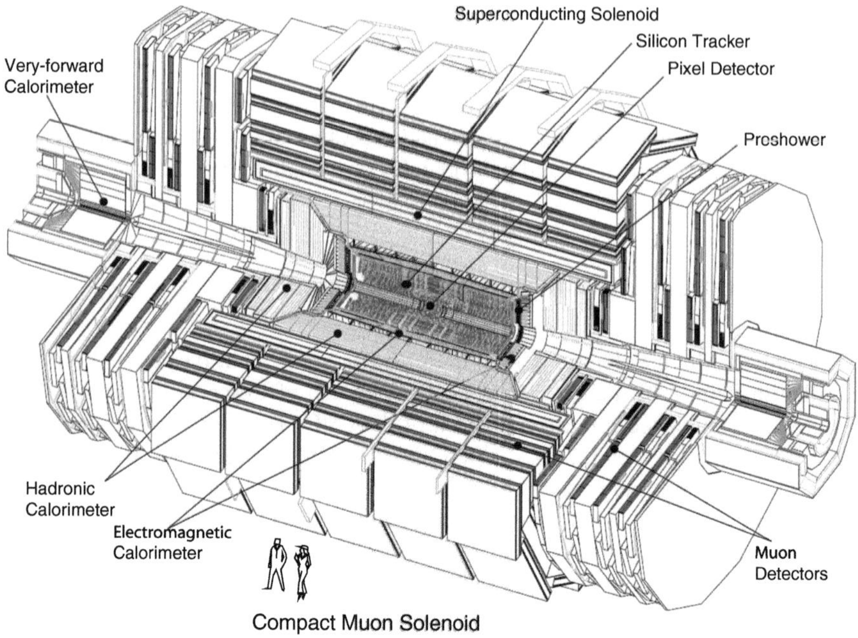

Fig. 3.5 Schematic of the CMS detector and its components

vertically upward and the z-axis pointing west towards the direction of the beam . The detector is cylindrically symmetric around the beam line. The azimuthal angle ϕ is defined with respect to to the x axis in the xy plane. The polar angle θ is measured from the z axis and the radial distance from the beam line is denoted as r. The pseudorapidity is defined as

$$\eta = -\ln \tan \frac{\theta}{2}, \tag{3.13}$$

and is preferred over the polar angle since particle production in minimum bias collisions is constant as a function of pseudorapidity. Since LHC is a hadron collider, the interesting observables (i.e energy and momentum) are defined as transverse to the beam by measuring their x and y components and they are denoted as E_{T} for the transverse energy and P_{T} for the transverse momentum.

3.2.2 Magnet

The choice of the magnet is crucial for ensuring good performance for a high energy physics experiment. Precise measurement of the charged particle momenta at a wide range of energies requires high bending power that can be achieved using strong magnetic fields. For charged particle in a uniform magnetic field, the momentum of the particle is given by

$$p = \gamma m v = q B r, \tag{3.14}$$

where q is the charge, m is the mass and r is the bending radius. The sagitta of the trajectory is given by

$$s = \frac{L^2}{8r} = \frac{q B L^2}{8p}, \tag{3.15}$$

where L is the path length in the magnetic field and assuming that the particle crosses the full solenoid L is equal to the radius of the solenoid. The transverse momentum resolution depends on the magnetic field and solenoid radius as

$$\frac{dp}{p} \propto \frac{p}{B L^2}, \tag{3.16}$$

therefore for improvement in the resolution both a high volume and a strong field is needed. CMS design targets both parameters since it utilizes a large solenoid of 6 m in diameter (and also a large tracker that defines the measurable path length) and a large magnetic field of 3.8 T. The CMS solenoid utilizes a 4-layer winding made from a stabilized reinforced Nb Ti conductor. The flux is returned by a 10,000 ton iron yoke comprised of five wheels and two endcaps that consist of three disks each. The yoke design allows for four muon stations to be installed with a very large geometric coverage.

3.2.3 Inner Tracking System

The tracker is the innermost sub-detector of the CMS apparatus. The purpose of the tracker is to reconstruct the trajectories and momenta of charged particles that emerge from LHC collision as well as the position of the primary and displaced vertices for rejecting additional interactions and to reconstruct objects with significant lifetimes such as b-jets. The inner tracking system consists of the pixel detector and the silicon strip tracker. For efficient reconstruction of LHC collisions, low hit occupancy and large hit redundancy are required. Low hit occupancy is established by high granularity of the detector and measurement redundancy is achieved by a large number of detector layers. Both those parameters result in a large number of on detector electronics which require efficient cooling and result in a large material budget in the tracker. Large material in the detector causes Coulomb scattering, limiting the track measurement accuracy, nuclear interactions between tracks and material, bremmstralung for electrons and conversions of photons to electron pairs. All those effects spoil the particle reconstruction resulting in a compromise in the tracker design. In addition, due to the huge particle flux the detector electronics must be designed to operate in a hard radiation environment, which implies additional design constraints.

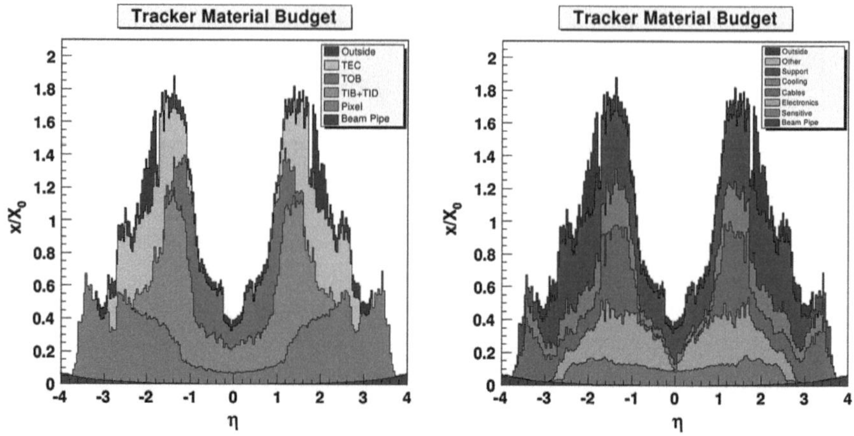

Fig. 3.6 Tracker material budget in units of radiation length as a function of pseudorapidity for the different sub-detectors (*left*) and by functional contributions (*right*)

CMS has opted for for a full-silicon tracking system that is composed of a pixel detector and a silicon strip tracker, providing high reconstruction performance up to $\eta < 2.4$. The dimensions are 5.4 m in length and 2.2 m in diameter, making the CMS tracker the largest inner detector ever built in a high energy physics experiment. Figure 3.6 shows the material budget of the tracker in units of radiation length as a function of pseudorapidity. The radiation length (X_0) corresponds to the mean distance over which a high energy electron loses all but $1/e$ of its energy or to $7/9$ of the mean free path for pair production of a high energy photon [3]. At $\eta = 0$, the tracker material budget corresponds to about $0.4\ X_0$ while at the boundary between barrel end endcap the material budget reaches a value of $1.8\ X_0$ due to cabling and other services in this region.

3.2.3.1 Pixel Detector

The pixel detector is the detector located nearest to the interaction point and faces very large particle flux in this region. It contributes precise tracking points in r, ϕ and z and therefore is responsible for a small impact parameter resolution that is important for good secondary vertex reconstruction of b and τ decays. To achieve this resolution, a pixel cell size of $100 \times 150\ \mu\text{m}^2$ is utilized. The pixel system has a zero-suppressed analog pulse height read-out scheme that improves position resolution and helps separating signal and noise hits as well as identifying large hit clusters from overlapping tracks.

The layout of the pixel system consists of three layers of pixel modules placed at radii of 4.4, 7.3 and 10.2 cm and two endcap discs extending from 6 to 15 cm in radius placed at 34.5 and 46.5cm from the nominal interaction point. The pseudorapidity range covered is $-2.5 < \eta < 2.5$ and the arrangement of the barrel layers and

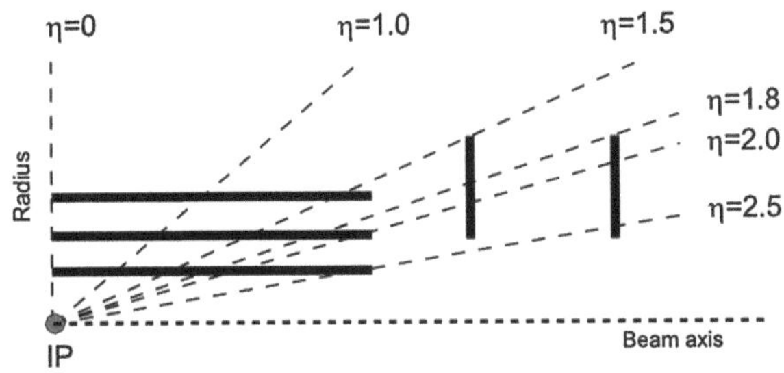

Fig. 3.7 Hit coverage of the CMS pixel detector

endcap disks ensures the existence of three measurement points almost over full covered range as shown in Fig. 3.7.

3.2.3.2 Silicon Strip Tracker

Outside the pixel detector the particle flux is much lower so silicon micro strip detectors are utilized. The silicon strip tracker (SST) consists of three different subsystems (Fig. 3.8). The Tracker Inner Barrel and Disks (TIB/TID), extending in a radius of 55 cm, are composed of four barrel layers and three disks at each end. The SST provides up to 4 $r - \phi$ measurements with a position resolution of 23 μm in the two inner layers and 35 μm in the two outer layers. The TIB and TID are surrounded by the Tracker Outer Barrel system (TOB). TOB consists of 6 barrel layers, providing a resolution of 53 μm in the first four layers and 35 μm for the two outer layers. It extends up to $z = 118$ cm and beyond this z range, the Tracker endcaps (TEC+, TEC−) provide the additional forward coverage up to $\eta < 2.5$. Each TEC provides 9 ϕ measurements per trajectory and extends to $z = 282$ cm. In addition the first two

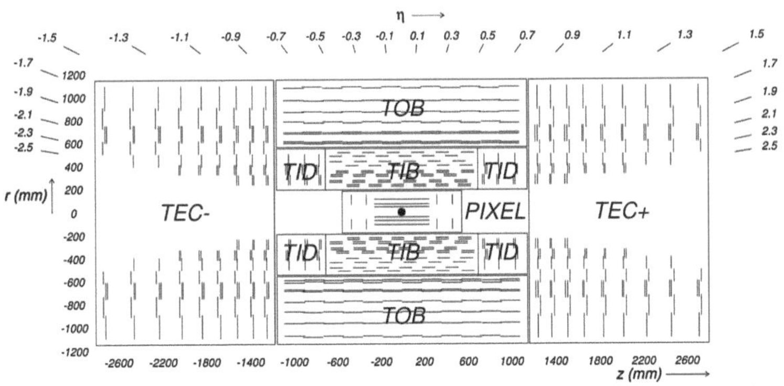

Fig. 3.8 Schematic cross section of the CMS silicon tracker. Each line corresponds to a detector module while double lines correspond to back-to back modules

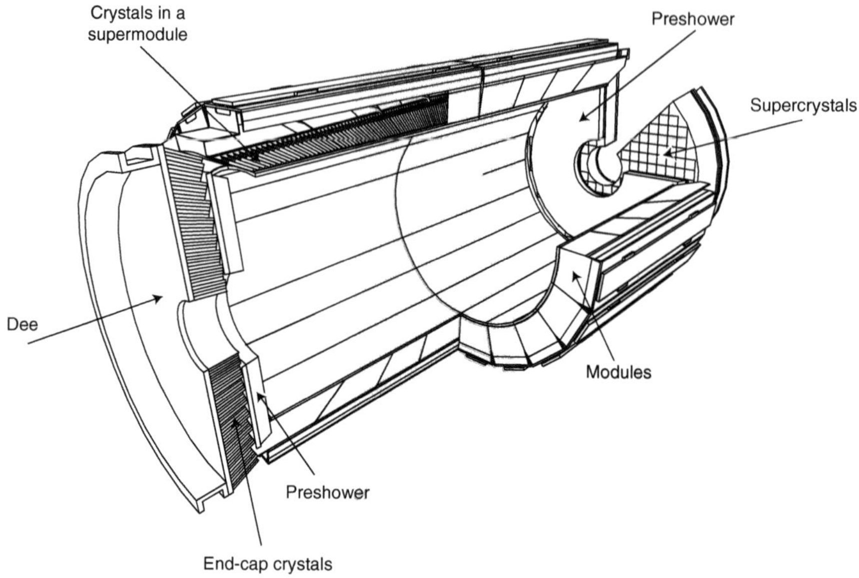

Fig. 3.9 Layout of the CMS Electromagnetic calorimeter showing the arrangement of crystals, supermodules and endcaps with the preshower in front

layers of TIB/TID and TOB, as well as rings 1, 2 and 5 of the TEC, carry a second micro-strip module with sensors in both sides with an angle of 100 mrad in order to provide a measurement of the second coordinate z or r in the barrel and discs respectively.

3.2.4 Electromagnetic Calorimeter

The CMS electromagnetic calorimeter (ECAL) is a hermetic homogeneous calorimeter made of 61200 lead tungstate ($PbWO_4$) crystals mounted in the central barrel part accompanied by 7324 crystals in each of the two endcaps. A preshower detector is installed at the face of each of the two endcaps. Avalanche photo-diodes (APDs) are used as photo-detectors in the barrel and vacuum photo-triodes (VPT) in the endcaps. The choice of high density crystals allowed CMS to build a compact calorimeter which provides fast response, high granularity and good energy resolution. The design of the electromagnetic calorimeter is motivated by the search for Higgs bosons in the di-photon final state where optimal position and energy resolution for photons is essential for Higgs discovery. The layout of the CMS ECAL is presented in Fig. 3.9.

3.2.4.1 Lead Tungstate Crystals

The choice of the crystals is very important for the performance of an electromagnetic calorimeter. For precise measurement of the energy, leakage outside the crystals has to be minimized. This is achieved by choosing a crystal material with low radiation length and moliere radius. Lead tungstate crystals provide a radiation length of 0.89 cm and their high density leads to a compact design resulting in 25.8 radiation lengths in 23 cm crystals. The position resolution is optimized by requiring small transverse size of the electromagnetic shower. The transverse profile of electromagnetic showers is expressed by the Moliere radius (R_M). About 90 % of the electromagnetic energy of a shower lies within a cylinder with radius equal to R_M. The Moliere radius of the lead tungstate crystals is 2.2 cm and combined with a barrel crystal front face of $2.2 \times 2.2\,\text{cm}^2$ ensures that the core of the shower will be included in a 2×2 crystal area. Lead tungstate provides optimal scintillation time since about 80 % of the light is emited in 25 ns which is the nominal beam crossing time separation.

3.2.4.2 Calorimeter Resolution

For energies of electromagnetic showers below 500 GeV, where energy leakage from the end of the calorimeter is negligible, the resolution of the electromagnetic calorimeter can be modeled in three terms:

$$\left(\frac{\sigma}{E}\right)^2 = \left(\frac{S}{\sqrt{E}}\right)^2 + \left(\frac{N}{E}\right)^2 + C^2, \tag{3.17}$$

where S is the stochastic term, N is the noise term and C is the constant term. The stochastic term describes statistics related fluctuations such as intrinsic shower fluctuations, photoelectron statistics, dead material in front of the calorimeter and sampling fluctuations. The main contribution to the constant term C are detector non-uniformity, calibration uncertainty, radiation damage to the calorimeter and leakage from the outside of a crystal. The constant term can be reduced by utilizing radiation hard media and performing in-situ calibration. CMS calorimeter accounts for both factors by utilizing a laser monitoring and calibration system. The noise term is related to the electronics noise and it is estimated by measuring the contribution from electronics noise after summing over some Moliere radii. One additional noise factor can come from pile-up where remnant energy from a previous crossing can be accounted in the measurement. Test beam results indicate that by measuring energy in a 3×3 crystal lattice, the resolution of the CMS ECAL is given by:

$$\left(\frac{\sigma}{E}\right)^2 = \left(\frac{2.8\,\%}{\sqrt{E}}\right)^2 + \left(\frac{0.12}{E}\right)^2 + (0.30\,\%)^2, \tag{3.18}$$

where E is the energy as measured in GeV.

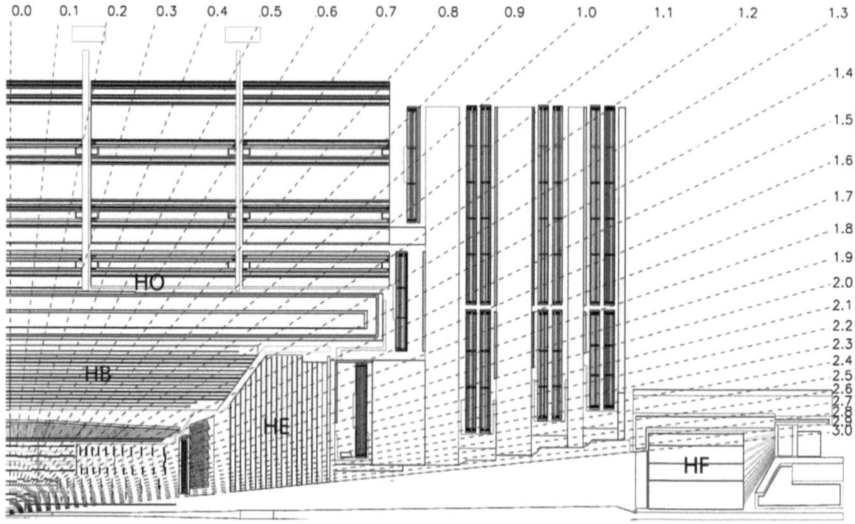

Fig. 3.10 Longitudinal view of the CMS detector showing the hadronic calorimeter components

3.2.5 Hadron Calorimeter

The CMS Hadron Calorimeter (HCAL) is responsible for measuring the energies of the hadrons produced in LHC collisions, especially the neutral ones since pions and kaons are measured very precisely by also estimating their momenta with the silicon tracker. In addition, for adequate measurement of the missing transverse energy, hermetic coverage is needed up to $|\eta| < 5$. The CMS HCAL is a brass/scintillator sampling calorimeter. Most of it is installed between the electromagnetic calorimeter and the solenoid magnet while small part of it is installed outside the magnet and it consists of four different subsystems:

- **Barrel Hadron Calorimeter (HB)**: located between ECAL barrel and the magnet covering up to $|\eta| < 1.4$.
- **Endcap Hadron Calorimeter (HE)**: located inside the magnet, consisting of two endcaps complementing HB and extending the coverage up to $|\eta| < 3.0$.
- **Outer Hadron Calorimeter (HO)**: located outside the magnet,in the barrel using the magnet as additional absorber providing the additional absorber length required for the barrel region.
- **Forward Hadron Calorimeter (HF)**: placed 11.2 m from the nominal interaction point extending the coverage from $|\eta| > 3$ up to $|\eta| < 5.2$.

The layout of the CMS HCAL is depicted in Fig. 3.10. The length scale of hadronic calorimetry is designated as the interaction length (λ_I) which corresponds to the mean free path of a hadron before undergoing an interaction in a given medium. The HB absorber consists of a 40 mm thick front steel plate, followed by eight 50.5 mm thick brass plates, six 56.5 mm thick brass plates and a 75 mm thick steel plate. The total absorber thickness at incident angle of 90° is 5.82 interaction lengths (λ_I) while at $\eta = 1.3$ it is 10.6 interaction lengths. The electromagnetic calorimeter adds about 1.1 λ_I of additional material. The endcap part uses the same absorber but the plates have a thickness of 79 mm. Between the absorber layers, plastic scintillator tiles as placed resulting in 70,000 tiles for the whole detector. The granularity of the tiles corresponds to $\Delta\eta \times \Delta\phi$ of 0.087×0.087 in the range of $|\eta| < 1.6$ and 0.17×0.17 in the region of $\eta > 1.6$. The HO utilizes the magnet as additional absorber corresponding to an additional $1.4\lambda_I$ of material at normal incidence and it consists of two scintillator layers with the same granularity as HB. The total depth in the central region sums to about $11.8\lambda_I$. The light produced in the HB scintillators is transferred to the Hybrid Photo Diodes (HPDs) via optical fibers. HPDs consist of a photo-cathode held at a HV -8 kV at a distance of approximately 3 mm from a pixelated silicon photo-diode and can provide a gain of 2000. The forward calorimeter (HF) design is very challenging, since in the LHC environment on average 760 GeV of energy are deposited in the forward calorimeters, compared to an average of 100 GeV in the rest of the detector. This flux introduces the need for radiation hard design. CMS opted for a Cherenkov based calorimeter using quartz fibers as an active material embedded in a steel absorber. The signal is generated when charged shower particles above the Cherenkov threshold generate light that is captured by photo-multipliers (PMT). The HF is therefore more sensitive to the electromagnetic showers and some relativistic charged pions. In the case of hadronic calorimetry, in contrast to the electromagnetic calorimetry, the best achievable energy resolution is limited due to the nature of the hadronic interactions. During a hadronic interaction, a part of the energy is purely electromagnetic due to the presence of π^0 and η mesons decaying to photon pairs and it is measured directly by the photo-detectors. Charged particles, on the other hand produce signal by ionization, excitation and nuclear interactions. In most of the cases, a significant fraction of the energy of the order of 20–35 %, deposited in a sampling calorimeter is not visible resulting in degraded resolution. For the CMS hadronic calorimeter, the resolution is described as:

$$\left(\frac{\sigma}{E}\right)^2 = \left(\frac{90\,\%}{\sqrt{E}}\right)^2 + (4.5\,\%)^2, \tag{3.19}$$

for the HB, HO and HE and

$$\left(\frac{\sigma}{E}\right)^2 = \left(\frac{172\,\%}{\sqrt{E}}\right)^2 + (9.0\,\%)^2, \tag{3.20}$$

for the HF.

3.2.6 Muon System

Muons are very important particles for LHC physics searches. Therefore, efficient and precise muon identification is one of the main requirements that influenced the design of CMS. Muons have a lifetime of about 2.19×10^{-6} s [3], even longer at relativistic speed. Therefore they cross the full detector before decaying. In addition, bremmstrahlung is suppressed by a factor of $(m_e/m_\mu)^2$ with respect to electrons for muons below 100 GeV which implies that the muon does not stop in the calorimeters but rather interacts via ionization. Muons are therefore expected to produce a track in the CMS tracker and then escape the calorimeter by depositing very small energy in it. In addition, for very energetic muons, the tracker cannot provide sufficient momentum resolution due to the small sagita of the very energetic track. This creates the need of a special muon system outside the calorimeters that can provide identification and momentum measurement. The purpose of the muon detectors is three-fold:

- Efficient muon identification
- Precise measurement of muon momentum and charge
- High speed able to provide muon triggering capabilities.

The muon detectors are placed inside the iron yoke to exploit the return field that provides bending of the particles for momentum measurement. The muon system follows a cylindrical geometry according to the whole CMS design and utilizes three types of gaseous detectors that are based on different complementary technologies. In the barrel region, drift tube (DT) chambers and resistive plate chambers (RPC) are used. The endcap region is instrumented by RPCs and also Cathode Strip Chambers (CSC) that provide better performance than DTs in a more dense radiation environment. A layout of the muon system is depicted in Fig. 3.11.

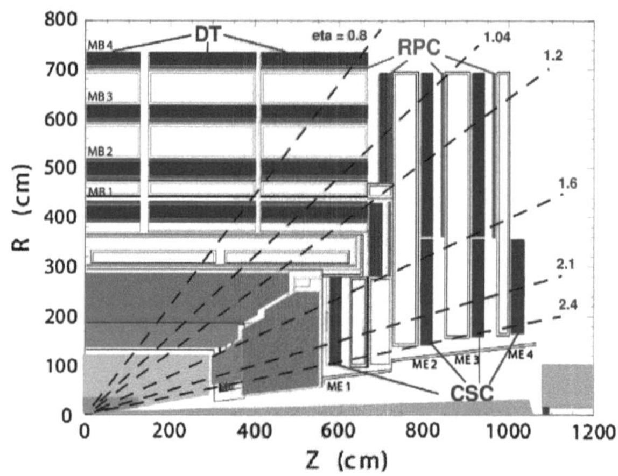

Fig. 3.11 Layout of one quadrant of the muon system in the $r-z$ plane showing the position of muon sub-detectors

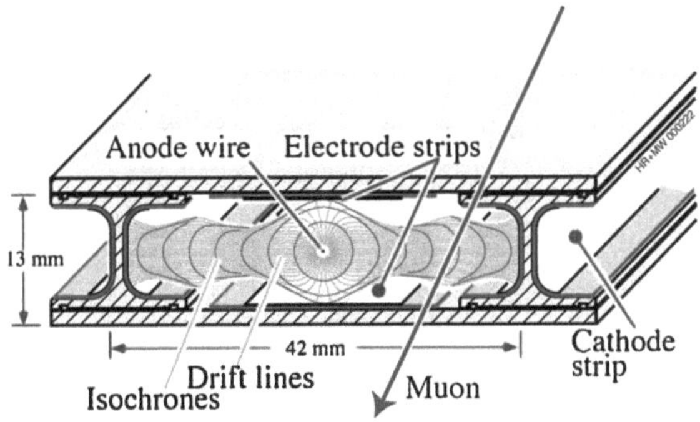

Fig. 3.12 The individual drift tube cell and operation principle

3.2.6.1 Drift Tube Chambers

In the CMS barrel, where the muon rate is under control, four layers of muon stations are used, occupied by Drift Tube (DT) Chambers covering up to $|\eta| < 1.2$. Drift Tube chambers consist of individual drift tube cells that contain a 50 μm-diameter anode wire and two electrode plates that create the drift electric field. The walls of the cell are grounded, acting as cathodes. The cells are filled with a gas mixture of 85 % Ar and 15 % CO_2 and the wire and electrodes are operated with a voltage difference of about 1.8 kV. Picking a transverse dimension for the cell of 21 mm to optimize drift time, gain and number of channels, all the above design conditions provide a gain of 10^5, resulting in a drift time of 380 ns and a linear relationship between drift time and drift path which is essential for the chamber to provide triggering capabilities. The basic DT cell is shown in Fig. 3.12. Each DT chamber is made of three (or two) super-layers (SL) where each super-layer is made of four layers of drift cells staggered by half a cell.The outer SLs have wires aligned parallel to the beam line providing a measurement in the transverse bending plane while in the inner SL, the wires are perpendicular to the beam line to provide a measurement of the z position of the track.

3.2.6.2 Cathode Strip Chambers

The Cathode Strip Chambers (CSCs) are installed in the endcaps, providing full muon coverage up to $|\eta| < 2.4$. The CSCs are multi-wire propotional chambers consisting of six planes of anode wires interleaved among seven cathode panels. Wires run azimuthially, defining the tracks radial component, while strips are milled on cathode panels and run lengthwise at a constant $\Delta\phi$ width. The angular ϕ position of the track is estimated by extrapolating the charge that is induced on the strips as shown in Fig. 3.13.

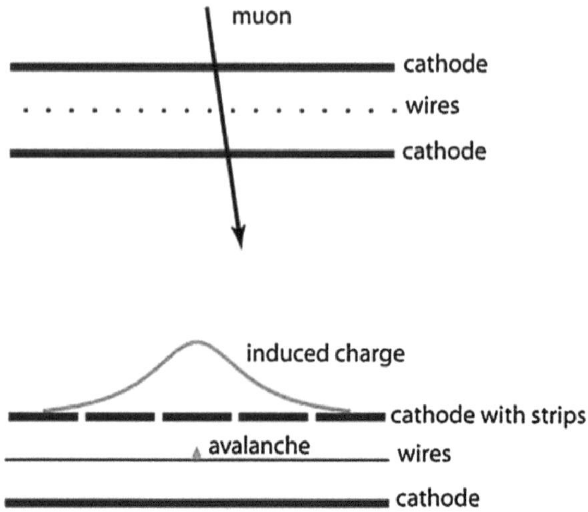

Fig. 3.13 Cathode strip chamber operation principle

The nominal gas mixture is 40 % Ar, 50 % CO_2 and 10 % CF_4. The addition of CF_4 is used to avoid polymerization of the wires. The CSCs can operate at high rates and in large and non-uniform magnetic fields without requiring precise monitoring of gas, pressure or temperature and can provide trigger and precision position measurement in the same device. The wires give very fast signals that provide very good time resolution while the development of the avalance on the strips gives very good position resolution.

3.2.6.3 Resistive Plate Chambers

Resistive Plate Chambers (RPC) are gaseous parallel plate detectors that combine adequate position resolution with very high speed. RPCs are able to tag the presence of an ionizing particle in much less than a bunch crossing, which makes RPCs an ideal trigger device since they can associate the correct bunch crossing with the muon. The CMS RPC chamber consists of two gaps operated in an avalanche mode with read-out strips in between. The total induced signal is the sum of the induced signal in both gaps. RPCs need intensive monitoring of temperature, humidity and pressure to ensure stability of conditions for proper operation. The RPC system spans both barrel and endcap and for the initial CMS operation chambers have been installed up to $\eta < 1.6$.

3.2.7 Trigger

The LHC provides proton collisions at high interaction rates with a designed bunch crossing rate of 40 MHz. Each recorded event in CMS has a nominal size of 0.5–1 MB which makes impossible to store and process this huge number of events. Therefore, a drastic rate reduction is achieved in real time using the trigger system. CMS uses a two-level trigger architecture that consists of the level 1 (L1) and high level trigger (HLT). The L1 trigger is based on hardware and consists of custom-designed, programmable electronics whereas the high level trigger is based on software that runs on a large filter farm of commercial processors.

3.2.7.1 L1 Trigger

The L1 Trigger uses coarse information from the Calorimeter and the Muon system, while holding the high resolution data in pipelined memories in the front-end electronics. The purpose of the L1 Trigger is to perform the sufficient reduction from the input crossing rate of 40 MHz to provide a maximum output rate of 100 kHz. The L1 Trigger starts from local information in calorimeters and muon detectors. This local information is encoded in the form of trigger primitives that define the coarse data to be used in the trigger algorithms, essentially hit patterns and segments in muon chambers and energies in the calorimeters. The next step uses the regional triggers that combine information from trigger primitives to build ranked trigger objects in localized regions of the detector. The Global Muon and Calorimeter Triggers then sort the regionally created candidates and forward the highest ranked ones to the Global Trigger (GT). No decision for selecting the event is taken before the Global Trigger. The GT combines all information from the global muon and calorimeter trigger and can flexibly apply topological requirements and requirements for combinations of objects which are crucial for efficient triggering of interesting physics events as luminosity increases. The different parts of the L1 Trigger system are shown in Fig. 3.14. The Calorimeter Trigger starts from the Trigger Primitive Generators (TPGs) in electromagnetic and hadron calorimeter. The Trigger primitives segmentation almost follows the HCAL segmentation of 0.087×0.087 towers in the barrel and larger in endcap therefore the individual ECAL crystal information is summed to similar sized towers for trigger primitive generators. However, since an electron or photon is expected to be very narrow in η direction and broader in the bending ϕ direction one additional bit is formed in ECAL TPGs requiring a deposit in a narrow crystal profile in η to be confined within two crystals in this direction. The transverse energy and the additional bits are then forwarded to the Regional Calorimeter Trigger (RCT). The RCT uses the towers to form electron or photon (e/γ) candidates and regions of 5×5 towers that are forwarded in the Global Calorimeter Trigger (GCT). In addition, the RCT creates sums of ECAL and HCAL energies to broader regions that are used later in the GCT to form jets. The regions are accompanied by a bit showing the compatibility of a region with the tau-lepton hypothesis. The GCT sorts

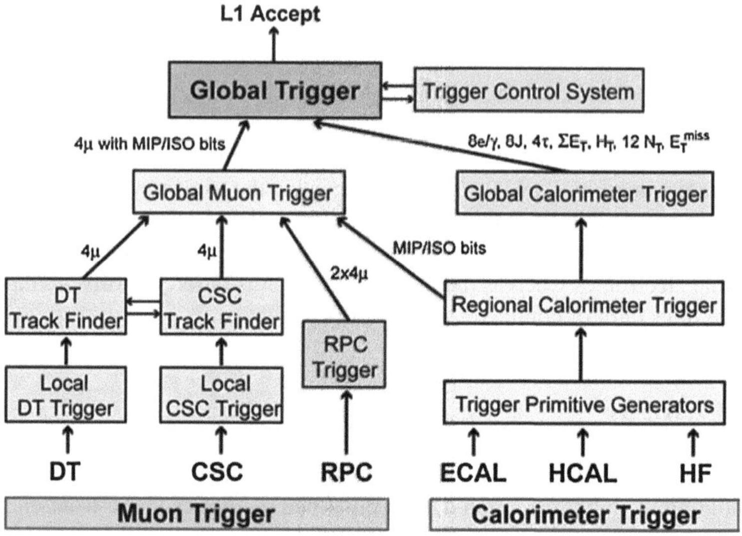

Fig. 3.14 Architecture of the CMS L1 trigger system

the electron candidates and forms jets by combining the regions created in the RCT. Then it sorts the highest e/γ and jets and forwards them in the Global Trigger.

The muon trigger uses all muon sub-detectors in the trigger. Trigger Primitives in the form of track segments are created in the DT and CSC system. Then in regional level the DT or the CSC track finders combine the segments to build muon tracks assigning a transverse momentum to them using Look-Up tables. The RPC system, being much faster provides its own track in the form of hit patterns created by the individual segments. All the regional information is forwarded to the global muon trigger (GMT) that combines all information, providing tracks with equal or better resolution than the regional ones. Then the tracks are ranked and forwarded in the Global Trigger.

3.2.7.2 Regional Calorimeter Trigger

The Regional Calorimeter Trigger consists of eighteen crates installed in nine racks in the CMS underground electronics room. Each crate corresponds to one slice in the (η, ϕ) Calorimeter TPG lattice and contains seven cards, with two regions per card resulting in fourteen regions per crate. For each trigger tower, the RCT takes as input two 8-bit calorimeter transverse energies, two energy characterization bits, an LHC bunch crossing bit and 5 bits of error detection from ECAL, HCAL and HF electronics. The RCT e/γ algorithm starts from the trigger towers and using look-up tables it forms the sum of the ECAL and HCAL energies as well as the ratio of HCAL to ECAL energy (H/E) which is used as a requirement to discriminate electrons and photons from pions and electromagnetic deposits in jets. The L1 e/γ algorithm is applied on trigger towers that have energy higher than their four immediate neighbors. The seeding tower must have a compatible electromagnetic deposit which is confirmed by requiring that the finegrain veto bit is not set. Then, the tower energy is combined with the nearest highest energy neighbor to account for particles that deposit energy into two towers. The sum of the two energies is associated with the e/γ candidate. To further discriminate against jets, isolation can be applied by requiring low energy surrounding the primary tower. If the eight towers around the primary tower have two full sides (L patterns) below a configurable threshold and those towers finegrain bit is not set,then the candidate is isolated. Each crate sorts the candidates and creates two collections of the four highest ranked isolated and non-isolated candidates. The isolated and non isolated collections are mutually exclusive.

The RCT also calculates the energy sums of the 4×4 tower regions that are used later for jet reconstruction in the GCT. During the creation of those regions, the RCT checks if the region is compatible with a hadronic τ lepton hypothesis by examining the energy profile. If the energy is not confined in a 2×2 tower cluster in the region where all other towers are below a programmable threshold then a tau veto bit is set indicating the region is not compatible with a tau lepton hypothesis. The transverse energy of the region and the tau veto bit is then sent to the GCT for jet reconstruction. The RCT e/γ and tau identification algorithms are depicted in Fig. 3.15. High performance of the Regional Calorimeter Trigger is essential for triggering electrons photons and taus that are crucial objects for the CMS physics program. The performance of the RCT is monitored during and after data-taking by constantly comparing the trigger output with the expected output from emulating the system. The RCT Emulator is a high level software package designed to provide the response of the hardware for any inputs. Using a high level design like this, it is possible to cross-check the hardware configuration and the observed behavior against the expected behavior of the system and spot possible problems that can appear in the cards that result in a non-expected behavior. The RCT data quality monitoring software package runs the RCT emulator on a subset of events and compares the emulator response with the hardware response for this event. Figure 3.16 shows a comparison of the e/γ candidate and region response in $\eta \times \phi$ bins of the L1 Trigger lattice showing excellent performance of the RCT in LHC physics runs.

3.2.7.3 High Level Trigger

The High Level Trigger (HLT) processes the events accepted by the L1 trigger by using a filter farm of commercial processors. The main goal of the HLT is to reduce the incoming event rate of 100 kHz by 10^3, reaching an output rate of the order of 300 Hz. After the event is selected at L1, the data corresponding to the same bunch crossing have to be transferred from the readout units to a single processor to be processed at the HLT. A fast network (Event Builder) is utilized that sequentially streams all the data of each crossing to an individual filter unit for HLT processing. Since HLT

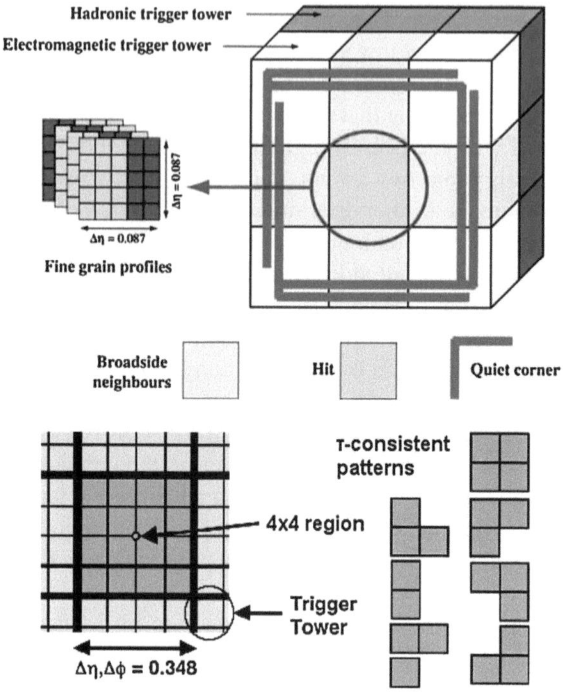

Fig. 3.15 Regional Calorimeter Trigger e/γ identification algorithm (*up*) and tau veto bit algorithm (*down*)

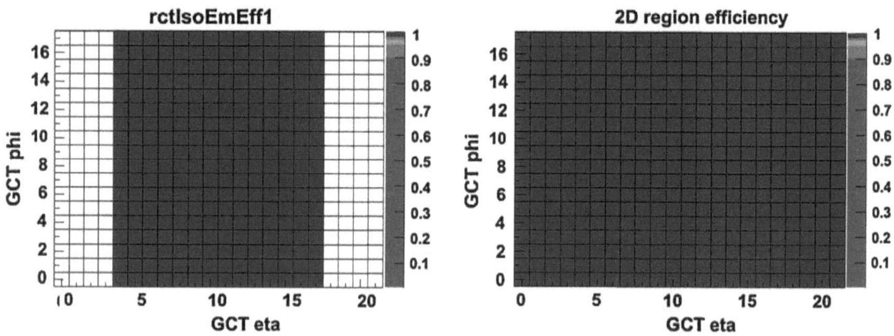

Fig. 3.16 RCT Data Quality Comparator between data and emulator of the transverse energy of e/γ and regions as a function of the tower index in $\eta - \phi$ space. Excellent agreement is observed between emulator and hardware

processing is much slower than L1, this procedure is serial, resulting each node in the farm processing a different event. The number of processors in the farm needs to be sufficient such that when the last node is occupied, the first node is available for a next event. However large safety factors have to be used for the case of failing nodes or nodes that take much more time to process events due to algorithm and events peculiarities (i.e huge number of tracks to be reconstructed). The time to process each event in the HLT mostly depends on the time to run specific complex algorithms such as track reconstruction. To optimize the time each node takes to process each event, the HLT utilizes several steps of event rejection. Simpler algorithms run first and basic requirements are applied before running more complex algorithms that take a longer time. If the event is rejected in the first step, complex algorithms do not run at all and the processing node is free to accept the next event. One example is the muon high level trigger that starts by unpacking the muon chamber digits. Regional reconstruction is performed in the muon system around the L1 track and if a good track exists, the next step unpacks the tracker digits and runs inner tracking around the same track making a full muon global track similar to the offline reconstruction. The algorithms used at the High Level Trigger are written in C++ and they are very similar to identical with the algorithms that run in the offline event reconstruction to achieve compatibility with the offline reconstruction and high trigger efficiency for offline selected high quality objects. More details about the High Level Trigger designed for this analysis is included in Chaps. 5 and 6.

References

1. L. Evans, P. Bryant, LHC machine. J. Instrum. **3**(08), S08001 (2008)
2. The CMS Collaboration et al. The CMS experiment at the CERN LHC. J. Instrum. **3**(08), S08004, (2008)
3. K. Nakamura, Particle data group. Review of particle physics. J. Phys. G: Nucl. Part. Phys. **37**(7A), 075021 (2010)

Chapter 4
Event Simulation

Physics and detector simulation is necessary in high energy physics to model complex physics processes, optimize analysis techniques and understand the performance of the numerous and complex sub-detectors of an apparatus such as CMS. Physics event and detector response simulation is performed using the Monte Carlo (MC) method [1]. The Monte Carlo method uses random sampling applied to a theoretical model to predict its expected behavior in realistic conditions. It relies on computer simulations and can give correct solutions especially in cases where a deterministic solution cannot be derived. Examples in high energy physics include event simulation where particles are produced in random direction and position but obey the theoretical constraints and also detector simulation where the detector behavior during passage of particles through it is modeled precisely including any external parameters such as electronics noise, noise of neighboring sub-detectors and other effects (e.g. temperature changes).

4.1 Physics Event Generation

Event generation is the first step in the simulation process and refers to the procedure of simulating a collision using programs known as MC generators [2] . Event generation is usually performed in many steps following a modular approach, often using different MC programs for each step.

4.1.1 Hard Scattering Process

The first step in event generation is the simulation of the hard scattering process. To illustrate the generation process we assume Z production at tree level and decay to tau

M. Bachtis, *Heavy Neutral Particle Decays to Tau Pairs*,
Springer Theses, DOI: 10.1007/978-3-319-03257-3_4,

pairs, $pp \to Z \to \tau^+\tau^-$. This process has a matrix element $\mathcal{M}(q\bar{q} \to Z \to \tau^+\tau^-)$ and the total cross section taking into account the parton distribution functions is:

$$d\sigma = \frac{1}{3} f_q f_{\bar{q}} \frac{1}{2(xys)^2} \left|\mathcal{M}(q\bar{q} \to Z \to \tau^+\tau^-)\right|^2 \frac{d\cos\theta d\phi dx dy}{8(2\pi)^2}, \tag{4.1}$$

where f_q, $f_{\bar{q}}$ are the parton distribution functions of the proton for each quark q and s is the square of the center of mass energy of the LHC, where $\sqrt{s} = 7$ TeV. The momentum fractions that each quark carries from the original proton are denoted as x and y. For the calculation of the cross section, Eq 4.1 is integrated using MC by picking uniform random numbers (θ, ϕ, x, y, q) and calculating the average $< d\sigma >$ which is an approximation of the cross section integral. Each element $d\sigma$ is referred to as the event weight.

In the case of event generation, the purpose is not the calculation of the cross section, but the creation of particles produced by a random collision due to the process under test, in this case Z production. To achieve this goal, the maximum event weight $d\sigma_{MAX}$ is calculated by scanning the phase space and then each infinitesimal element $d\sigma$ is normalized by calculating $d\sigma/d\sigma_{MAX}$. Then, a uniformly distributed number $0 < g < 1$ is generated and if $d\sigma/d\sigma_{MAX} > g$ this element in phase space is generated and the kinematics of the decay are obtained. Before the output is converted to particles that will be simulated within the detector, additional processes, e.g. NLO effects, (radiation of gluons, quarks and photons), evolution of jets (hadronization) and decays of tau leptons need to be performed.

4.1.2 Parton Shower, Underlying Event and Hadronization

After the hard process has been generated, higher order effects are added by "evolving" the event using parton shower simulation, which allows partons to split in pairs of other partons (i.e $g \to q\bar{q}$, $q \to gq$). The resulting partons can branch further, resulting in a large number of final quarks and gluons that are not allowed to exist in isolation due to color confinement. The next step is the modeling of the underlying event (UE). UE refers to the beam remnants of the proton after the partons contributing to the hard scatter have been pulled out. The beam remnants recoil against a very small transverse momentum, present due to the motion of the partons inside the proton. The resulting partons are then grouped together into color singlet composite hadrons using a phenomenological model known as hadronization. The hadronization scale is in the non-pertubative regime. Therefore, crude phenomenological models which contain several parameters tuned with experimental data are used. Because the hadronization scale is much smaller than the hard scales, the impact of the choice of the hadronization model is small for most physics processes. Any resonances created during the hadronization process are decayed. Finally the possibility of multi-parton interactions where two or more partons per proton can

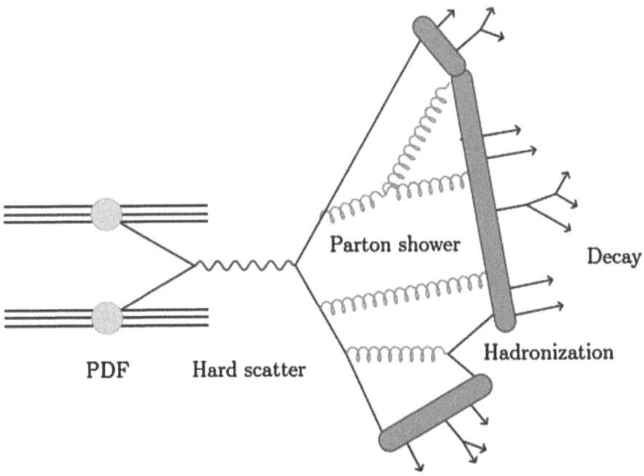

Fig. 4.1 Illustration of the hard scattering process, parton shower, hadronization and decay during the generation of an event with two quarks in the final state

interact is taken into account and is added to the simulated event. Figure 4.1 shows the different steps in the event generation using as an example a proton collision resulting in two quarks followed by parton shower and hadronization.

4.1.3 MC Generator Programs

Several MC simulation programs are used to model the processes in this analysis depending on the peculiarities of the different final states under study. A brief description of the respective generators is described in the next sections.

4.1.3.1 PYTHIA

PYTHIA [3] is a general purpose event generator for lepton and hadron colliders. It consists of a sub-process library containing about 240 different $2 \to n$ sub-processes all at leading order. Standard processes include SM vector boson and vector boson pair production, QCD processes, SUSY, Higgs production and exotic physics. Initial and final state showers are added to provide more realistic configurations especially for the internal structure of jets. Hadronization is also performed by using a phenomenological model knows as the "Lund string model" [3] and the underlying event is simulated by $2 \to 2$ scatterings between the remaining partons. The Parton shower and hadronization developed in PYTHIA is also used when it is interfaced to other generators such as MADGRAPH and POWHEG.

4.1.3.2 POWHEG

The POWHEG method [4] is an improvement to PYTHIA since it utilizes NLO calculations combined with parton shower. The main idea of the POWHEG method is that the hardest emission (the one with the highest p_T) is simulated according to the exact NLO cross section. Then during the parton shower, the hardest emission is excluded and subsequent emissions are vetoed if they are harder than the hardest emission. The POWHEG method, compared with PYTHIA, provides a much better description of basic processes such as vector boson and Higgs production and is used explicitly to describe final states with low multiplicity such as inclusive W/Z production.

4.1.3.3 MADGRAPH

PYTHIA is very effective when describing simple $2 \rightarrow 2$ processes. However, in most experimental cases there are additional hard particles in the final state. One example is the $W+$jets background in the $H/Z \rightarrow \tau\tau$ analysis where one electron or muon is produced from the W decay and one jet is faking a hadronic τ. This example involves the production of a lepton, a neutrino and a jet and can be complicated further by considering the example of vector boson fusion Higgs production and decay to tau pairs, $qqH \rightarrow qq\tau\tau$. In this final state, W+3 jets is a major background with one jet faking a tau and two other jets are faking the VBF signature. The MADGRAPH generator [5] provides a better description for those types of final states utilizing calculations of diagrams of multiple final state objects at tree level. Given a user process, MADGRAPH automatically generates the amplitudes for all relevant sub-processes and produces the mappings for the iteration over the phase space. Once the events are generated, MADGRAPH is interfaced to PYTHIA for the parton shower and hadronization procedures.

4.1.3.4 TAUOLA

TAUOLA [6] is a MC package used explicitly to model the decays of the tau leptons including proper description of the tau polarization. The description of the tau polarization is very important in view of the fact that due to the scalar nature of the Higgs boson, the helicity states of the produced tau leptons are different than the ones from $Z \rightarrow \tau\tau$ decay and this needs to be described properly. TAUOLA utilizes an individual phase space generator for each decay mode including specific modeling of the weak and hadronic current and it can be interfaced to other generators that produce the hard scattering process, e.g. PYTHIA, POWHEG and MADGRAPH.

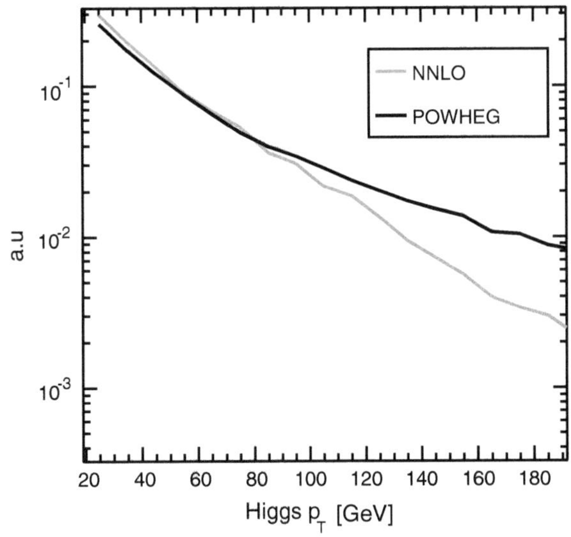

Fig. 4.2 Transverse momentum of the Higgs boson as modeled by POWHEG and NNLO for a Higgs with mass of 120 GeV

4.1.4 K-factors

Even if event generators are becoming more and more sophisticated, there are usually several calculations performed in higher orders that affect observable quantities in the experiment. To account for those additional effects, the specific observables of a generator such as POWHEG are re-weighted to match the spectra of the observables of the higher order calculation. The weights that are produced via this process are referred as K-factors. One example is the re-weighting of the p_T of the Higgs boson to match the spectrum obtained by NNLO calculations. In this case the NNLO spectrum is obtained using the FeHiPro [7] cross section integrator and its spectrum is used to correct the POWHEG spectrum.

Figure 4.2 shows the transverse momentum of the Higgs boson as modeled by POWHEG and NNLO using FeHiPro. POWHEG overestimates the cross section for high transverse momenta of the Higgs. Therefore the POWHEG sample is re-weighted according to the above distributions to match the NNLO spectrum.

4.2 Detector Simulation

The complexity of the CMS Detector requires very sophisticated simulation to properly reproduce the detector behavior in the presence of particles from proton collisions. The detailed simulation is performed using the GEANT4 [8] toolkit. GEANT4 relies on the accurate detector description including the full geometry, the materials of the detecting devices and the dead material (e.g. cables, support, cooling) to simulate the particle response. It takes as input particles from the event generator and then propagates them through the detector taking into account the measured magnetic field map (for charged particles) and any interactions between particle and material such as bremmstralung, multiple scattering and photon conversions. At the final stage, GEANT4 produces a set of simulated hits in the active material such as the energy loss of a particle in the sensitive volume of a detector. The simulated hits are then fed to emulators of the readout and trigger electronics to simulate the event and provide it in digitized format which is equivalent to the digital information produced by real data events. All subsequent stages, such as event reconstruction, use the same input collection and are the same when running on simulated events or data.

4.2.1 Simulation of Multiple Interactions

The number of interactions per crossing can very significantly affect the performance of identifying interesting events. Therefore, the effects of pileup need to be simulated properly to account for the changes in performance and allow for studies to improve the performance for operation on higher luminosities. Pileup is simulated by overlaying events at the hit level. The hard interaction is superimposed with additional simulated minimum bias events by combining the hits between them. Both in-time (multiple vertices per event) and out-of-time (detector response remnants from previous crossings) are simulated. For out-of-time pileup, the time information is shifted by one bunch crossing allowing the slower detectors to be affected by the information in the other crossings. Along with the simulated hits, the generator information is also provided to allow for studies of the merged event.

The distribution of pile-up interactions in the data depends on the run period and the respective LHC configuration. However, the production of the simulated samples is performed centrally before or during the run. To avoid multiple simulation production runs, the simulated samples are produced with a flat distribution of PU interactions and that distribution is re-weighted to match the distribution of interactions in data.

4.3 Simulated Samples

The complete set of the samples and the MC generators used for this analysis is presented in Table 4.1. The SM Higgs signal is modeled using POWHEG samples. For the gluon fusion process the POWHEG spectrum is re-weighted to match the spectrum predicted by NNLO. In the MSSM case, where the Higgs is produced in association with b quarks, the PYTHIA simulation is preferred since it provides an accurate description for the $2 \rightarrow 2$ process that is produced. For the $Z \rightarrow \tau\tau$

Table 4.1 List of simulated samples used in this analysis including normalization cross sections and preselection requirements in generator stage

Sample	Generator	$\sigma \times \epsilon[pb]$	Comments
SM $gg \to H \to \tau\tau$	POWHEG	-	with NNLO k-factors
SM $qqH \to qq\tau\tau$	POWHEG	-	-
SM $VH \to V\tau\tau$	POWHEG	-	V denotes vector boson or $t\bar{t}$ pair
SUSY $gg \to H \to \tau\tau$	POWHEG	-	-
SUSY $bbH \to bb\tau\tau$	PYTHIA	-	-
$Z \to \tau\tau$	POWHEG	1667	gen. $m_{\ell\ell} > 20\,\text{GeV}$
$Z \to \ell\ell$+jets	MADGRAPH	3048	gen. $m_{\ell\ell} > 50\,\text{GeV}$
$W \to \ell\nu$+jets	MADGRAPH	31314	-
$t\bar{t}$+jets	MADGRAPH	158	-
Single t (s channel)	POWHEG	4.21	-
Single t (t channel)	POWHEG	64.6	-
Single t (tW channel)	POWHEG	10.6	-
WW	PYTHIA	43	-
WZ	PYTHIA	18.2	-
ZZ	PYTHIA	5.9	-
QCD μ enriched	PYTHIA		$\hat{p_T} > 20\,\text{GeV}$, $\mu\ p_T > 15\,\text{GeV}$
QCD EM enriched	PYTHIA		$\hat{p_T} > 30\,\text{GeV}$
QCD $b/c \to e\nu$	PYTHIA		$\hat{p_T} > 30\,\text{GeV}$

production, POWHEG is used to describe the inclusive production while MADGRAPH samples are used to describe Z+jets production in the case of final states with jets (e.g. $\tau\tau + b$, $\tau\tau$+VBF). W boson production is modeled by MADGRAPH since there is always at least one jet in the final state that will fake hadronic decays of tau lepton in addition to the prompt lepton from the W decay. Top pair production is simulated using MADGRAPH and includes cases of the presence of additional jets produced with the top pair. Single top production is a very small background and is modeled using POWHEG including all three contributions from t, s and tW channels. Diboson samples simulated with PYTHIA correspond to a small fraction of the background. In the case of QCD multijet production, estimation of shapes and normalization in the analysis is performed using data. However, QCD samples were used for some analysis optimization purposes. For final states with muons, a QCD sample with a requirement of the presence of a muon with $p_T > 15\,\text{GeV}$ and $\eta < 2.5$ at the generator level is used. This sample covers the cases where a muon is produced by a heavy flavor decay or decays in flight of π and K mesons. While most of the muons in QCD events are real muons, in the case of electrons, the QCD background contribution is dominated by fake electrons from jets with electromagnetic fluctuations. This background is modeled by requiring an electromagnetic signature at the generator level. The electromagnetic signature requires an isolated sum of energy from electrons or photons of 20 GeV in a cone of $\Delta R = 0.1$. The energy deposited by hadrons in this cone should be less than 50 % of the EM energy. In addition, less than 5 GeV of energy coming from charged particles and less than 10 GeV of energy from neutral particles is allowed to exist in an annulus defined by an outer cone of $\Delta R = 0.2$. Finally the EM enriched sample is required to not contain heavy flavor quark decays to electrons. For the heavy flavor case, a separate sample is produced. For each event in a simulated sample an event weight is assigned as following:

$$w_{\text{gen}} = \frac{\sigma \times \epsilon \times L}{N_{\text{gen}}}, \tag{4.2}$$

where σ is the cross section, ϵ is the efficiency of the generator level requirements, L is the integrated luminosity and N_{gen} is the number of produced events before any generator level requirements.

References

1. R. Eckhardt, S. Ulam, J. Von Neumann, The Monte Carlo method. Los Alamos science, Special Issue **15**(15), 131–137 (1987)
2. M.A. Dobbs et al. Les houches guidebook to monte carlo generators for hadron collider, physics. hep-ph/0403045 (2004)
3. T. Sjostrand, S. Mrenna, P. Skands. PYTHIA 6.4 physics and manual. JHEP **05**(026), (2006)
4. S. Alioli et al., A general framework for implementing nlo calculations in shower monte carlo programs: the powheg box. JHEP **1006**, 043 (2010)
5. Fabio Maltoni, Tim Stelzer, MadEvent: automatic event generation with MadGraph. J. High Energy Phys. **2003**(02), 027–027 (2003)
6. Z. Was, TAUOLA the library for tau lepton decay. Nucl. Phys. Proc. Suppl. **98**, 96–102 (2001)
7. Charalampos Anastasiou, Kirill Melnikov, Higgs boson production at hadron colliders in NNLO QCD. Nucl. Phys. **B646**, 220–256 (2002)
8. S. Agostinelli et al., G4-a simulation toolkit. Nucl. Instr. Methods Phys. Res. Sect. A **506**(3), 250–303 (2003)

Chapter 5
Event Reconstruction

The collected collision information from all sub-detectors is stored in 'raw' data format that can be used for online event reconstruction (in HLT) or offline event reconstruction for data analysis. The reconstruction process utilizes several algorithms to identify particle candidates and higher level quantities used to identify and analyze useful events.

5.1 Track and Vertex Reconstruction

Tracking and vertexing are the most important parts of CMS reconstruction since identification of charged particles is crucial for the CMS Physics programme. In addition efficient reconstruction of the primary event vertex is extremely important to separate the hard scattering process from pileup interactions. This section describes the track and vertex reconstruction in CMS.

5.1.1 Track Reconstruction

Track Reconstruction in CMS [1] starts from the clustering of pixel and silicon tracker signals into "hits" produced by charged particles on the silicon detectors. The hit position and their uncertainties are estimated and then used for track seed generators. Track seeding is the process of creating initial trajectories using a small number of hits. To define a trajectory, at least three hits or two hits and a beam/vertex constraint are used. Seeds are created in the inner part of the tracker where the occupancy per unit area and the effects due to the material budget (i.e. nuclear interactions) are minimal. Several seeding categories are used, including pixel triplets, pixel pairs and vertex/beamspot constraint, pixel-strip pairs with vertex or beamspot constraint or strip pairs with beamspot constraint. After the seeds are created, tracks are found by propagating the seed trajectory to each detector layer using a Kalman filter approach [2]. At each new layer, compatible hits are associated with the track

M. Bachtis, *Heavy Neutral Particle Decays to Tau Pairs*,
Springer Theses, DOI: 10.1007/978-3-319-03257-3_5,

and the track parameters are updated. In case of multiple compatible hits, additional track candidates are created. The track finding algorithm also accounts for the case where a track did not leave a hit in the corresponding detector layer before propagating to the next layer. During the iterative reconstruction process, a lot of ambiguities are created in the case of multiple seeds per track or multiple tracks per seed. Trajectories are cleaned to reduce those ambiguities by examining the number of shared hits between track pairs. If the number of shared hits is larger than 50 % of the hits of the track with the lower number of hits, the track with lower number of hits is discarded. In the case of equal number of hits, the track with the highest χ^2 is discarded. After ambiguities have been removed, a final track fit is performed using a Kalman Filter with smoothing [2]. During the final fit, spurious hits that have very large residuals from the trajectory are also removed. Finally to reject fake tracks, especially in the case of high multiplicity in jets, several tracking quality criteria, e.g. the number of hits, normalized χ^2 and vertex compatibility are used.

5.1.2 Iterative Tracking

Iterative tracking is tuned to maintain high efficiency and low fake rate even for tracks reconstructed in a very dense environment (i.e. inside jets) which is used extensively for reconstructing charged particles in the Particle Flow reconstruction described in the following sections. Iterative tracking starts with seed and track reconstruction using very tight criteria leading to moderate efficiency but very low fake rate. Then hits are unambiguously removed from very high quality tracks and used to seed tracks in subsequent iterations with loosened seeding criteria. This increases the tracking efficiency while keeping the fake rate low due to the fact that the pruned hits have reduced combinatorics. Full track reconstruction and cleaning is performed before the next iteration. Iterative tracking proceeds in six steps. In the first three steps 99.5 % of the tracks associated with isolated muons and 90 % of the pion tracks are already reconstructed. The last iterations use also the silicon tracker for seeding tracks created far from the interaction point to improve conversion and nuclear interaction reconstruction.

5.1.3 Vertex Reconstruction

Vertex Reconstruction [3] uses the Deterministic Annealing (DA) algorithm [4]. The problem of vertex finding refers to the association of a set of reconstructed tracks with a number of reconstructed vertices along the beam line. If z_i is the position of tracks extrapolated to the beam line and z_k the position of the vertex, a figure of merit can be defined as:

$$\chi^2 = \sum_{i,k} c_{ik} \frac{(z_i - z_k)^2}{\sigma_i^2}, \tag{5.1}$$

where c_{ik} corresponds to a matrix defining if a track is associated with the vertex ($c_{ik} = 1$) or not ($c_{ik} = 0$). The solution of the vertex finding problem is the minimization of this χ^2. However the DA algorithm uses a different probabilistic approach where the assignment matrix c_{ik} corresponds to probabilities that can take any value $0 < c_{ik} < 1$. Therefore, instead of minimizing the χ^2, the algorithm finds the most probable assignment for a given χ^2 and then the χ^2 is gradually reduced defining a new assignment in each iteration. For a vertex to be selected in the analysis, it is required to have a z position of smaller than 24 cm with respect to the center of the detector, a ρ position of smaller than 2 cm with respect to the nominal interaction point and more than four degrees of freedom. Out of the vertices selected with the above criteria, the event primary vertex is defined as the one that maximizes:

$$\sum_i p_{\mathrm{T}}^2, \qquad (5.2)$$

where p_{T} is the transverse momenta of the associated tracks.

5.2 Electron Reconstruction and Identification

Electrons in CMS are reconstructed using either ECAL or tracker seeds. ECAL seeded reconstruction starts from an electromagnetic deposit in the ECAL and matches this deposit with a track to build an electron candidate [5]. Track seeded reconstruction starts from tracks and associates Bremsstrahlung radiation to reconstruct electron objects [6, 7].

5.2.1 ECAL Seeded Reconstruction

The first step of the ECAL reconstruction is clustering of the calorimeter hits into ECAL clusters. Basic clusters are created from ECAL cells. In the case of the endcap ($1.479 < \eta < 3.0$) the ECAL cluster positions are extrapolated to the preshower detector (ES) and ES clusters are built. The total cluster energy in the endcap region is the sum of the endcap ECAL and preshower cluster energies.

The presence of material in front of the calorimeter ($0.4 - 1.8X_0$) can cause Bremsstrahlung of electrons and subsequent conversions of radiated photons. Because of the strong magnetic field of CMS, the energy flow associated with primary electrons or converted photons spreads in the ϕ direction with increasing distances from the collision vertex within the tracker volume. To account for those additional deposits when calculating the electron energy, basic clusters are summed into superclusters which are extended in ϕ direction. Superclusters with $E_{\mathrm{T}} > 4$ GeV are selected and matched to track seeds in the inner tracker layers to built electron tracks from those seeds. Trajectories are reconstructed using a dedicated model of the electron energy loss and fitted using a Gaussian Sum Filter (GSF) algorithm [8]. Finally the momentum assignment for the electron is performed by combining the energy measured in the ECAL with the momentum assigned to the track by the GSF algorithm.

Based on the desired performance, additional selection requirements can be applied on the supercluster and on the ECAL-track matching variables. The basic discriminating variables for an ECAL seeded electron are:

- The transverse shower profile along the η direction ($\sigma_{\eta\eta}$).
- The difference in pseudorapidity ($\Delta\eta$) and azimuthal angle ($\Delta\phi$) between the energy weighted position of the supercluster and the GSF track as calculated in the position of the primary vertex.
- The longitudinal shower profile quantified by the ratio of hadronic calorimeter energy deposited in the supercluster area divided by the energy of the supercluster (H/E).
- The Bremsstrahlung ratio (f_{brem}) defined as the momentum difference in the outer and inner measurement state divided by the inner state momentum.
- The ECAL-track compatibility(E/p), defined as the ratio of the supercluster energy divided by the momentum of the GSF track.
- The transverse impact parameter between the track and the primary vertex (d_{xy}).

5.2.2 Track Seeded Reconstruction

In the case of low energy electrons or electrons produced inside jets, the calorimeter signature is very broad due to the strong magnetic field, resulting in mis-measurement of the electron energy. In the case of electrons in jets, pollution of the supercluster energy by particles produced near the electron also degrades the energy measurement. To account for those effects, a dedicated track seeded electron algorithm has been developed. The algorithm starts from all reconstructed tracks and electromagnetic clusters. Starting from the inner track position, the Bremsstrahlung hypothesis is tested at each detector layer by checking the presence of an ECAL cluster at the point on the calorimeter surface calculated by extrapolating a straight line from the track position and momentum vector at the corresponding detector layer. The process is repeated for all layers and a supercluster is defined by summing all linked electromagnetic cluster deposits. This procedure is very efficient but does not account for the cases of additional showering effects produced by conversions of the radiated photons that broaden the energy deposits and cannot be approximated using the tangent line method. Those effects are recovered in two ways. Converted photon reconstruction is performed using the track pairs to recover some of the conversions and create the photon four vector. In the case of isolated electrons, the ECAL driven supercluster footprint is used to assign the full energy to the electron. Due to the high probability of reconstructing a fake low energy electron or electron in jets, a multivariate approach is used to discriminate electrons from pions exploiting a Boosted Decision Tree (BDT) that is described in detail elsewhere [6, 7].

5.2.3 Rejection of Electrons from Converted Photons

An important background contribution to di-tau analysis, where one tau decays to an electron and one hadronically, is due to $\gamma+$ jets production where a high energy photon converts to an electron–positron pair and a jet is misidentified as a hadronic tau. To reject this background, it is essential to identify and veto electrons produced by photon conversions. Photon conversions [9] are reconstructed by combining opposite sign track pairs (produced with the Kalman or GSF algorithm) and performing a vertex fit of those tracks to identify conversions. The distance between the conversion vertex and the primary vertex and the vertex fit probability as calculated from the fit are used to discriminate against fake conversions. In addition, to reject electrons coming from conversions, electron tracks are required to have the expected number of hits in the pixel detector given the track position and direction, accounting for dead channels that are not expected to give hits in the pixel.

5.2.4 Electron Identification

Tight electron identification is necessary to reject large QCD backgrounds in the di-tau analysis. Electron identification requirements are tuned separately in the barrel and the endcap as well as in different electron transverse momentum ranges. In the analysis of the electron+muon final state, it is desirable to identify electrons with transverse energy just above 10 GeV but below 20 GeV where the background is much larger therefore a different set of criteria are applied. The electron identification requirements are presented in detail in Table 5.1.

Table 5.1 Electron identification requirements tuned separately for high and low p_T electrons and barrel and endcap detector regions

Electron ID requirements				
Requirement	$p_T < 20\,\text{GeV/c}$		$p_T > 20\,\text{GeV/c}$	
	Barrel	Endcap	Barrel	Endcap
$\sigma_{i\eta i\eta} <$	0.01	0.03	0.01	0.03
$\|\Delta\eta_{in}\| <$	0.004	0.005	0.004	0.007
$\|\Delta\phi_{in}\| <$	0.03	0.02	0.03	0.03
$H/E <$	0.025	0.025	0.04	0.025
$f_{brem} >$ or $\eta < 1.0$ and $E/p >$	0.15 or 0.95	0.15 or 0.95	–	–
Missing inner tracker hits	0	0	0	0
Vertex fit probability <	10^{-6}	10^{-6}	10^{-6}	10^{-6}
Vertex distance from PV >	2 cm	2 cm	2 cm	2 cm

5.2.5 Electron Trigger

The electron trigger starts from identified e/γ candidates produced at L1 Trigger. Electron objects are created in Regional Calorimeter Trigger where basic identification requirements are imposed in the electron compatibility ($H/E < 0.05$). The four most energetic e/γ objects are forwarded to the Global trigger where a L1 decision takes place. In the case that a L1 Trigger path that involves electromagnetic objects fires the trigger, all four highest e/γ objects are used as seeds for the electron HLT. In the first step of the electron HLT, regional ECAL reconstruction is performed around the L1 objects using full granularity at crystal level. ECAL hits are clustered to clusters and superclusters using identical configuration with the offline reconstruction. Basic electromagnetic compatibility criteria (shower shape, H/E) and calorimetric isolation are applied as well as thresholds on the electron object. For the events that pass this step, pixel reconstruction is performed to create track seeds. Track seeds are required to match the supercluster candidate. Then seeds that are found are extrapolated to the supercluster and several track-supercluster matching identification criteria are applied to define a HLT electron object. In the case of isolated triggers, track based isolation is applied as final step to trigger the event.

5.3 Muon Reconstruction and Identification

Muons are reconstructed using combined information from the tracker and the muon detectors. The first step of the muon reconstruction corresponds to separate track reconstruction in the tracker and in the muon system. Based on this information three muon reconstruction approaches are used: standalone-muons, global muons and tracker muons [10].

5.3.1 Standalone Muon Reconstruction

Standalone muon reconstruction is performed using only tracks from the muon system. Standalone muon tracks are reconstructed using the Kalman Filter technique [2] starting from local track segment seeds or from the trajectory parameters estimated by the L1 Trigger electronics. Tracks are then propagated using an iterative algorithm taking into account the magnetic field and multiple scattering in the material. The choice of including additional hits from each station is based on χ^2 compatibility trials during each propagation step. After basic identification criteria, the tracks are propagated to the closest approach of the beam-line. A beamspot constraint is applied to improve the track p_T resolution. Standalone muon tracks are mainly used to reconstruct global muons.

5.3.2 Global Muon Reconstruction

Global Muon reconstruction starts from matching a standalone muon track to a track in the inner tracker. The large multiplicity of tracks in the tracker necessitates the selection of a subset that roughly corresponds in momentum and position to the standalone muon tracks. The matching is performed in two steps. Firstly, tracks in a rectangular $\eta \times \phi$ region are selected around the standalone muon tracks. Secondly, an iteration is performed on the inner tracks applying more stringent spatial and momentum matching criteria to select the best matching track. Both inner and standalone tracks under test are propagated to a common surface and a quality test is performed comparing the position, direction and momentum compatibility on the propagated plane and the primary vertex. After selecting the inner tracks that correspond to the standalone muon tracks, the next step is the creation of a global track that uses all tracker and muon hits. There is no additional pattern recognition at this step but the hits from both sub-detectors are refitted together to form a global track. In the case of more than one inner track candidate, the track pair that gives the best χ^2 is selected resulting in only one inner track per standalone muon track. The momentum resolution for muons with $p_T < 200$ GeV/c is driven by the excellent tracker resolution provided by the inner measurement. However at higher transverse momenta, the standalone track provides significant improvement to this measurement. However in $H/Z \rightarrow \tau\tau$ analysis, higher transverse momenta are not relevant since muons from tau decays tend to be soft.

5.3.3 Tracker Muon Reconstruction

Global muon reconstruction relies on the presence of a high quality standalone track in the muon system. In the case of low p_T muons, the muon system information is minimal and the standalone muon reconstruction fails. The tracker muon reconstruction considers silicon tracker tracks as muons if they are compatible with signatures in the muon system. The tracker muon algorithm starts from tracks at a given p or p_T and extrapolates them to the calorimeter and the muon system accounting for material effects. The algorithm then searches for compatible segments in the muon system. The distance and the pull between the extrapolated track and the muon segment is used as quality criteria for the track-segment match. During tracker muon reconstruction, a global refit is not performed. Therefore the momentum of the muon corresponds to the momentum of the inner track. In the case that this muon is also a global muon the momentum associated with it is given by the global refit. In the case of tracker muon reconstruction where many tracks are near a set of segments in the muon system, it is quite possible that the same set of segments is associated to more than one track. To define a 1–1 association between segments and tracks, two algorithms are implemented for arbitration exploiting the minimum ΔX or minimum $\Delta R = \Delta X^2 + \Delta Y^2$ between segment and track. Tracker muons are very effective for lower p_T muon reconstruction exploiting the excellent silicon track resolution to extrapolate precisely into the muon system.

5.3.4 Muon Identification

In most the cases high quality muons are reconstructed both as tracker and global muons. The main source of background to isolated muons from tau decays are muons coming from heavy flavor (b/c) production and muons produced in K/π decays in flight. Muons produced in heavy flavor jets can be suppressed by exploiting the large lifetime of B hadrons by requiring a small impact parameter, which is defined as the distance of closest approach between the track and the primary vertex. In addition, requirement of a minimum number of pixel hits constrains the track to be produced near the collision point. In the case of decays in flight, the χ^2 of the global track is used to suppress them when a significant kink is produced due to the neutrino recoil in the $K/\pi \rightarrow \mu\nu_\mu$ decay.

Since the background from heavy flavor and decays in flight is very significant in the $Z/H \rightarrow \tau\tau$ analysis, a tight selection is required at the muon level. Muons are required to be identified both as global and tracker muons. The global muon track is required to have at least one good muon hit and the global track $\chi 2/\mathrm{ndof}$ must be smaller than 10. The inner track is required to have at least 10 hits in the silicon tracker and at least one hit in the pixel detector. The tracker muon is required to have passed segment ΔX arbitration and to have at least two matched segments. Finally the relative transverse momentum resolution $\Delta p_T/p_T$ is required to be better than 10 %, where Δp_T is the error of the momentum measurement. Additional selection based on the impact parameter is applied depending on the analyzed tau pair final state. For the final states consisting of a lepton and a hadronic tau, the transverse impact parameter with respect to the primary vertex is required to be $\Delta_{xy} < 0.045$ cm; large enough to account for the tau lepton lifetime.

The muon performance is quantified using muons produced in Z boson decays. Figure 5.1 shows the reconstructed muon transverse momentum and di-muon invariant mass in simulation and 4.7 fb^{-1} of CMS data collected in 2011. Despite the very good agreement, residual differences in the muon identification efficiency are quantified in Chap. 7 where the simulation is corrected to match the muon performance in data.

5.3.5 Muon Trigger

The muon trigger [11] is essential for all aspects of the CMS physics program including analysis with tau leptons where one tau decays to muon and two neutrinos. The muon HLT starts from the output of the L1 muon trigger. The L1 muon trigger provides coarse information about the muon candidate momenta and position,

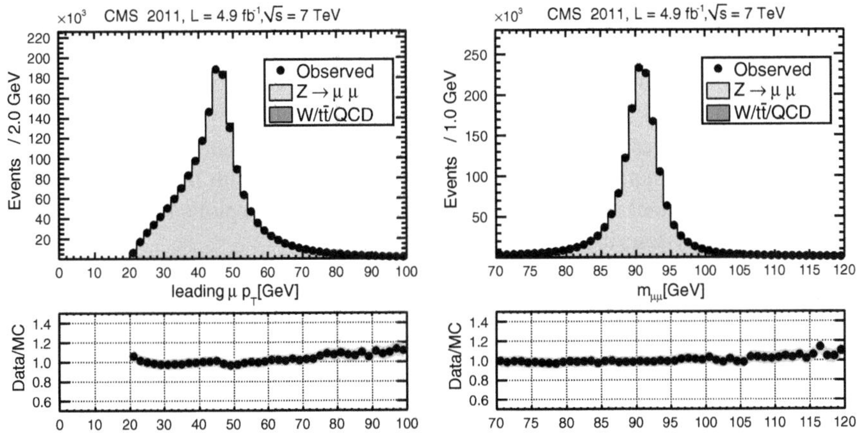

Fig. 5.1 Leading muon transverse momentum (*left*) and di-muon invariant mass (*right*) for the selected $Z \rightarrow \mu\mu$ sample. The filled band in the ratio corresponds to the uncertainties due to statistics and luminosity measurement (2.2 %). No corrections are applied to the simulation

associates muons to respective bunch crossings and provides event timing information. The subsequent muon HLT provides full track reconstruction, muon isolation and is used as first step in many multiple object triggers such as muon+tau trigger. The muon HLT starts from the four highest quality candidates produced by the L1 Global Muon Trigger (GMT). Standalone muon reconstruction is performed in the regions defined by those candidates and using calorimeter information, isolation deposits are defined in the case of the isolated muon triggers. Specific thresholds on the transverse momenta and isolation are applied at this stage depending on the trigger requirements. Muon candidates that pass those requirements are forwarded to the next step that involves track reconstruction. Full silicon track reconstruction is applied in a constrained region defined by the standalone track and the vertex and a refit is performed between the tracker and the standalone tracks to form a global muon track. Finally in the case of isolated muon triggers, track isolation can be applied around the muon to define an isolated muon candidate. Details of the efficiency of the muon reconstruction and trigger using data is described in Chap. 7.

5.4 Particle Flow Reconstruction

Particle Flow (PF) reconstruction [6] combines information of all CMS sub-detectors to define a unique event description in the form of mutually exclusive particles, the PF candidates. PF candidates are then used to build higher level objects such as jets, taus and missing transverse energy as well as isolation deposits for leptons. Most stable particles produced in proton collisions at the LHC have a rather low p_T even in collisions involving large momentum transfer. For example, during the hadronization of a quark or gluon jet with $p_{\mathrm{T}} > 500\,\mathrm{GeV/c}$, the average p_{T} carried by each constituent is of the order of $10\,\mathrm{GeV/c}$ while for jets with $p_{\mathrm{T}} > 100\,\mathrm{GeV/c}$ the average momentum per constituent reduces to few GeV/c. The goal of the PF algorithm is to identify as many particles as possible even if they have low transverse momenta and energies. To achieve this goal, specific techniques are implemented in tracking and calorimeter clustering to identify candidates with high efficiency and low fake rate. The track reconstruction in the PF algorithm exploits fully the iterative tracking approach described in Sect. 5.1.2.

5.4.1 Calorimeter Clustering

The purpose of the calorimeter clustering is to identify neutral stable particles and also deposits from charged hadrons and electrons. In case of charged hadrons with low quality reconstructed tracks, calorimeter measurement improves the energy/momentum measurement. Calorimeter clustering is performed separately in each sub-detector: ECAL barrel, ECAL endcap, HCAL barrel, HCAL endcap, Preshower first and second layer. In the case of forward calorimeter, clustering is not performed but each hit gives rise to an individual cluster. The clustering algorithm proceeds in three steps. First, cluster seeds are created using local calorimeter cell energy maxima. Second, topological clusters are created by the seeds adding hits that have an adjacent crystal side with the seed, requiring that the cluster energy is above a threshold defined by the noise in the calorimeter electronics, namely 80 MeV in the ECAL barrel, 300 MeV in the ECAL endcap and 800 MeV in the HCAL. Third, each topological cluster is used as a seed for many particle flow clusters.

5.4.2 Link Algorithm

A given particle produced in CMS is expected to give rise to several individual PF "elements" in different sub-detectors. For example, a charged hadron will give rise to a track and a calorimeter deposit. Since the main purpose of Particle Flow is to reconstruct the individual particles and remove the double counting of those deposits, a linking algorithm is necessary. The link algorithm is performed on all individual pairs of elements defining a linking distance to quantify the quality of the link. The CMS detector is ideal for PF reconstruction because the high granularity of the tracking and calorimeter which enables resolution necessary to link the specific detector elements making the separation between particles easier. The linking between a track and calorimeter deposits proceeds as follows: the track is extrapolated from its last measurement hit in the tracker to the two layers of the Preshower Detector, the ECAL and HCAL. The extrapolation points in ECAL correspond to a depth that is defined by the maximum of a typical longitudinal EM shower profile, while the HCAL extrapolation point corresponds to a depth of one interaction length which is typical for a hadron shower. The track and cluster are linked if the extrapolation point is within the cluster boundaries. In the case of electrons, the track-ECAL linking is performed using the track seeded electron reconstruction described in Sect. 5.2.2. In the case of a pair of clusters the link is established if the cluster position in the more granular calorimeter (ECAL/PS) is within the cluster envelope of the less granular calorimeter (HCAL/ECAL).

5.4.3 Particle Reconstruction

The particle reconstruction and identification is performed on the blocks of linked elements. First electrons are reconstructed using the track seeded approach that was developed to be efficient on different types of electrons including those inside jets. All the clusters linked by the Bremsstrahlung recovery or the PF supercluster are associated with the electron and removed from further processing. Then muons that are not inside jets are reconstructed by Particle Flow if they are reconstructed as global muons. For muons inside jets, a tighter approach is used due to the higher fake rate from pion and kaon decays. Some additional muons inside jets not passing the tight selection are recovered later in the step of calorimeter-track association during the charged hadron reconstruction. After muons and electrons have been identified, the next step is identification of charged hadrons, neutral hadrons and photons. Tracks can be directly connected to a number of ECAL and HCAL clusters. Neutral particles are identified after the calorimeter energy associated to charged particles has been removed. For the charged hadron hypothesis, energy in ECAL and HCAL is compared to the track momentum using calibrated cluster energy, taking into account the corresponding uncertainties. In the case of multiple tracks linked to clusters, the sum of the momenta are required to correspond to the linked calorimetric energy, while in the case of several clusters linked to the track, the closest one is kept for the comparison. Special attention is given when a charged hadron overlaps with a photon, where we have hadronic energy and many ECAL clusters linked to it. The total calorimeter energy in this case is defined as the sum of the HCAL energy and sum of the ECAL energy computed cluster by cluster starting from the one with the smallest link distance until the total calorimeter energy is equal to the track momentum. For the case that the calorimeter energy is in agreement with the track momentum a combined fit is performed using the measurements and the uncertainties to redefine the particle four vector. In the case that the total linked calorimeter energy is significantly higher than the track momentum, taking into account the calorimeter resolution, the farthest clusters are treated as separate neutral particles. If the excess of energy is larger than the total ECAL energy both photon and a neutral hadrons are created otherwise the excess gives rise to only a photon. The precedence given in ECAL to photons over neutral hadrons is justified by the observation that, in jets, 25 % of the jet energy is carried by photons while neutral hadrons leave only 3 % of the jet energy in the ECAL. On the other hand, there is the case that the calorimeter energy is much smaller than the track momentum. In this case a search for muons

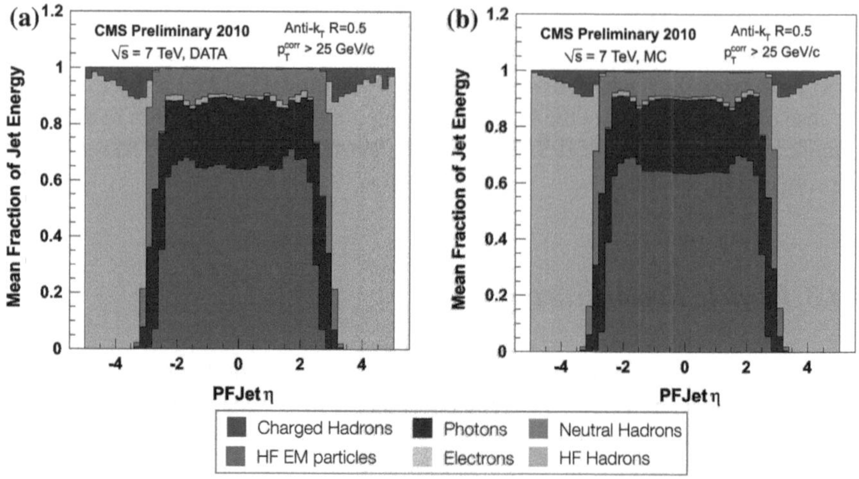

Fig. 5.2 Reconstructed jet energy fractions as a function of pseudorapidity **a** in the data and **b** in the simulation. From *bottom* to *top* in the central region: charged hadrons, photons, electrons, and neutral hadrons. In the forward regions: hadronic deposits, electromagnetic deposits [12]

not identified by the tight selection is performed by checking the compatibility of the muon hypothesis with the calorimeter energy. All remaining tracks are treated as charged pions. Figure 5.2 shows the reconstructed particle content of jets clustered using PF candidates for data and simulation. Beyond the tracker acceptance ($\eta > 2.5$) only neutral particles are reconstructed while at the forward region the algorithm is restricted to HF hits. Excellent agreement is observed for the particle content in data and simulation.

5.5 Lepton Isolation

Electrons and muons produced by tau decays are expected to be isolated in the detector while leptons from heavy flavor decays and π/K decays in flight are expected to be inside jets. Lepton isolation is usually calculated using detector information from the tracker and the calorimeters. This sub-detector based isolation is defined by summing individual sub detector deposits in a cone around the lepton:

$$I_\ell^{Det} = \frac{\sum p_{\mathrm{T}}^{\mathrm{track}} + \sum E_{\mathrm{T}}^{\mathrm{ECAL}} + \sum E_{\mathrm{T}}^{\mathrm{HCAL}}}{p_{\mathrm{T}}^{l}}, \tag{5.3}$$

where p_{T}^{track} is the sum of the transverse momenta of the tracks, E_{T}^{ECAL} is the transverse energy sum of ECAL hits, E_{T}^{HCAL} is defined as the transverse energy sum of HCAL hits and p_{T}^{l} is the transverse momentum of the lepton. One characteristic of the sub-detector based isolation is that calorimeter deposits of charged particles such as pions and kaons are double counted since the calorimeter and tracker deposits are both added in the sum. To avoid this double-counting, particle based isolation is defined using objects created by the PF algorithm. All charged particles are considered in the isolation sum, while photons and neutral hadrons are required to have $E_T > 0.5$ GeV. The isolation variables sum up particles of the above types in a cone of $\Delta R = 0.4$ around the lepton axis. The particle based isolation is defined using the following formula:

$$I_l^{PF} = \frac{\sum p_{\mathrm{T}}^{\mathrm{charged}} + \sum E_{\mathrm{T}}^{\gamma} + \sum E_{\mathrm{T}}^{\mathrm{neutral}}}{p_{\mathrm{T}}^{l}}, \tag{5.4}$$

where $\sum p_{\mathrm{T}}^{charged}$ is the sum of the transverse momenta of charged particles, $\sum E_{\mathrm{T}}^{\gamma}$ is the sum of the transverse energies of the PF Photons and $\sum E_{\mathrm{T}}^{neutral}$, the sum of the transverse energy of PF neutral hadrons. Isolation is studied in data using a $Z \rightarrow \mu\mu$ sample selected by requiring two high quality muons (as defined in Sect. 5.3) with $p_{\mathrm{T}} > 20\,\mathrm{GeV}/c$ that have opposite charges. The most energetic one is required to be isolated using $I_l^{PF} < 0.5$ and the invariant mass of the two muons is required to be within the Z mass window ($70 < m_{\mu\mu} < 120$). The subleading muon is used as a 'probe' to quantify the isolation performance. In a similar way, a heavy flavor sample is selected by requiring the most energetic muon to be inside jets ($I_l^{PF} > 0.5$) and the mass window to be outside the Z peak ($m_{\mu\mu} < 70$ or $m_{\mu\mu} > 120$). Heavy flavor production is very important for this analysis since production of b/c quarks is the major QCD contribution.

Figure 5.3 shows a Receiver Operator Characteristic (ROC) curve of the muon isolation efficiency from muons from Z production versus the efficiency for muons from heavy flavor semileptonic decays. The curve is created by scanning different cut values and measuring the reduction in the Z enriched and heavy flavor enriched

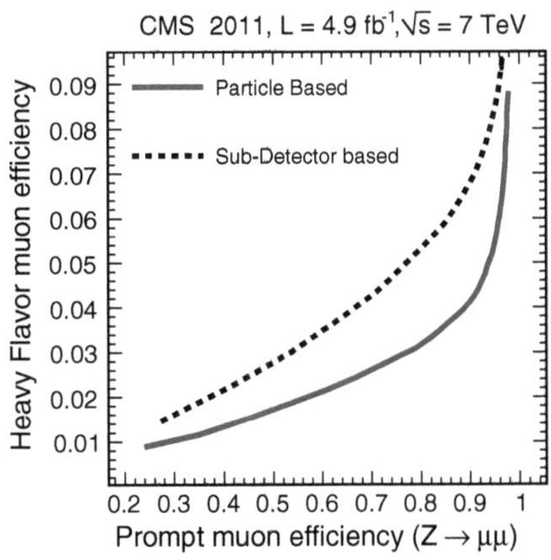

Fig. 5.3 Receiver operator characteristic (ROC) curves of the muon efficiency from muons from Z versus the efficiency for muons from heavy flavor production. Particle based reconstruction shows improved performance with respect to the sub-detector based one

sample and illustrates the discriminator power of each isolation method. Particle based isolation demonstrates better discrimination power and for a Z muon efficiency of 90 % the heavy flavor acceptance is suppressed by about a factor of 2 with respect to the subdetector based isolation.

5.5.1 Isolation Optimization at Higher Luminosities

In the case of high instantaneous luminosity, the number of primary interactions in the detector becomes significant and particles from other interactions degrade the isolation performance. In the case of charged particles, the vertex information can be used to remove charged particles from other interactions. The association probabilities of the deterministic annealing algorithm as described in Sect. 5.1.3 are used to associate charged particles to the respective vertices. In case charged particles are not associated with any vertex by the vertex algorithm, they are associated with the vertices by their closest distance in the z axis after they are extrapolated to the beam line. Using this separation algorithm, only charged particles from the primary vertex are included in the isolation sum described in Eq. (5.4). The vertex association accounts for roughly two thirds of the particles corresponding to the charged fraction of a proton collision but still does not account for neutral particles produced by pile-up interactions. For this purpose, a specific correction is implemented to account for the neutral energy from other interactions known as the $\Delta\beta$ correction. The $\Delta\beta$ correction makes use of the Particle based isolation to predict the neutral particle deposits in the isolation cone based on particle deposits due to pile-up in the same cone. A charged particle transverse momentum sum is created by summing all over the charged particles inside the isolation cone of the lepton but by requiring that those charged particles do not originate from the primary vertex. Then the charged particle sum is converted into an expected neutral deposit by the simple assumption that the average charged to neutral particle ratio is 2:1. The particle based isolation sum is modified as follows:

$$I_{\ell}^{PF/\Delta\beta} = \frac{\sum p_{\mathrm{T}}^{\mathrm{charged}} + \max\left(\sum E_{\mathrm{T}}^{\gamma} + \sum E_{\mathrm{T}}^{neutral} - 0.5 \sum E_{\mathrm{T}}^{\mathrm{charged}}(PU), 0.0\right)}{p_{\mathrm{T}}^{l}}, \tag{5.5}$$

where $\sum E_{\mathrm{T}}^{\mathrm{charged}}(PU)$ is the charged particle p_{T} sum when charged particles are not associated with the primary vertex. The average sum of the charged particle transverse momenta from pileup interactions is presented in Fig. 5.4 as a function of the number of reconstructed primary vertices and the muon pseudorapidity. There is a linear relation between the average sum and the vertices with a slope of 0.23 GeV/vertex corresponding to about 0.12 GeV of neutral energy in the cone of $\Delta R = 0.4$ per reconstructed vertex. The average pileup isolation transverse energy shows a structure as a function of muon pseudorapidity. There is a bump in the highest material region around $\eta = 1.5$ that corresponds to the increased number of photon conversions and

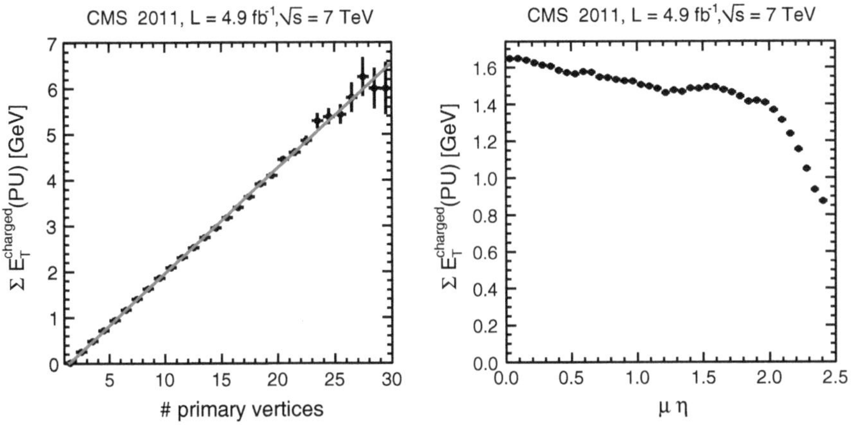

Fig. 5.4 Average isolation sums using charged particles from PU vertices as a function of the number of reconstructed vertices (*left*) and muon pseudorapidity (*right*) for a $Z \rightarrow \mu\mu$ sample

nuclear interactions in this region creating more charged particles. This excess is visible in the particle content inside the jets presented in Fig. 5.2. In addition there is a decrease of the average pileup isolation transverse energy at the end of the tracker acceptance where there are less charged particles reconstructed. Due to those detector effects the correction is expected to be higher in the high material region and lower in the end of the tracker acceptance. The muons in this analysis are restricted to $\eta < 2.1$, and therefore the performance in the forward part of the tracker is not very relevant.

The efficiency of the particle based isolation is measured in the $Z \rightarrow \mu\mu$ and the heavy flavor samples, for an isolation requirement of $I_{\ell}^{PF} < 0.15$. Figure 5.5 shows the isolation efficiency as a function of the number of reconstructed primary vertices

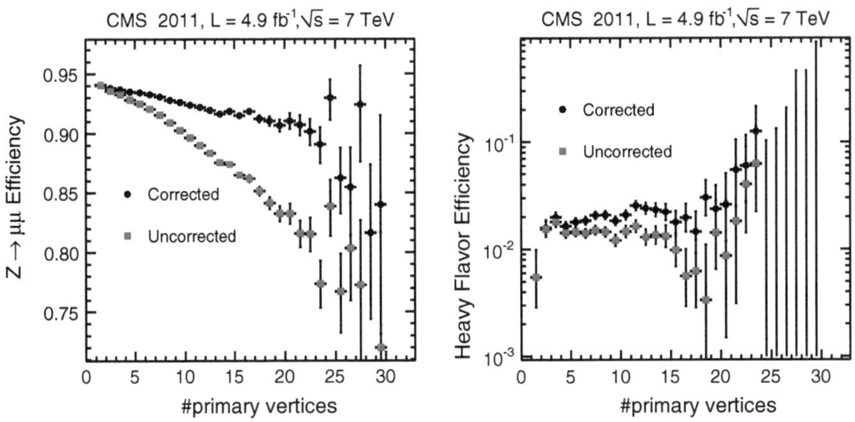

Fig. 5.5 Efficiency for $Z \rightarrow \mu\mu$ (*left*) and heavy flavor (*right*) enriched samples for the isolation requirement $I_{\ell} < 0.15$ with and without $\Delta\beta$ corrections applied

for the two samples. PU corrected isolation shows a much smaller degradation of the $Z \rightarrow \mu\mu$ efficiency as the number of vertices increase. In the case of the heavy flavor sample the efficiency is higher for the corrected isolation-resulting in higher heavy flavor contamination- but the performance is approximately flat as a function of the number of vertices.

5.6 Jet and Missing Transverse Energy Reconstruction

Jets are very important objects since they are the signature of hadronizing quarks. Efficient jet reconstruction is crucial to identify final states with jets such as associated production of Higgs with b quarks or vector boson fusion production of the Standard Model Higgs. Missing transverse energy is a very important measure of the presence of invisible particles in the detector which is very important to identify $W \rightarrow l\nu$ and tau decays.

5.6.1 Jet Reconstruction

Jets are reconstructed at particle level by clustering particles produced by the Particle Flow Reconstruction. The jet reconstruction using PF resembles much better the properties at generator/true level than traditional calorimeter hit jet reconstruction, therefore the comparison with theory is much more efficient. Roughly, the 90 % of the jet energy which corresponds to charged hadrons and photons is reconstructed with excellent precision by the PF algorithm while the remaining 10 % corresponds to neutral hadrons that are affected by the poor HCAL resolution, especially for low neutral hadron E_{T}.

Particles are clustered into jets using the 'anti-k_T' algorithm [13]. An important requirement for this jet clustering algorithm is to be infrared and collinear (IRC) safe. The algorithm is infrared safe if it is robust against the presence of soft particles radiated by the partons (e.g. gluon radiation). The algorithm is collinear safe if parton splits into a pair of collinear partons are recombined to the original one. The IRC-safe property of the jet algorithm is important for comparisons of the jets to theoretical calculations of any perturbative order. The anti-k_T algorithm uses the effective distances:

$$d_{ij} = \min\left(\frac{1}{k_{T,i}^2}, \frac{1}{k_{T,j}^2}\right)\frac{\Delta_{ij}^2}{R^2}, \tag{5.6a}$$

$$d_{iB} = \frac{1}{k_{T,i}^2}, \tag{5.6b}$$

where $\Delta_{ij}^2 = (y_i - y_j)^2 + (\phi_i - \phi_j)^2$ and $k_{T,i}$, y_i and ϕ_i are the transverse momentum, rapidity and azimuth of the particle i respectively and R is referred as the algorithm parameter. The distance d_{ij} is denoted as the distance between two entities (particles or pseudo-jets) and d_{iB} is denoted as the distance between a particle and the beam. The algorithm iterates in steps starting from the highest p_{T} entity and clustering softer particles around it. For each clustering step the two distances are calculated and if the smaller one is d_{ij}, the particle is added in the jet while if it is d_{iB}, a new jet is created.

The effect of the anti-k_T algorithm is that softer particles tend to cluster with harder ones before clustering with other soft particles. A hard particle with no nearby neighbors will have all the soft particles within R added to it, resulting in a conical jet. If another hard particle is between $R < \Delta_{ij} < 2R$, the two hard particles will split the surrounding softer particles, weighted according to their relative transverse momenta, resulting in only partly conical jets. Two hard particles within $\Delta_{ij} < R$ will combine into one jet, the center weighted by their relative momenta to the harder particle and the shape formed from the union of the two cones around the particles plus the cone around the final jet. One strength of this reconstruction method is that softer particles do not modify the shape of the jet, yet it is still flexible for harder radiation. An example of the anti-k_T algorithm applied to a parton level event with added soft radiation is shown in Fig. 5.6. The jets used in this analysis have a parameter R of 0.5.

5.6.2 Jet Energy Corrections

The UE and pile-up can smear the energy of the jets as well as change the distribution of particles inside jets. Jets are corrected for those effects using a pile-up subtraction using jet areas [14] called L1 Fastjet correction. Assuming that a jet has an area in rapidity-azimuth coordinates denoted by A and the mean contribution from

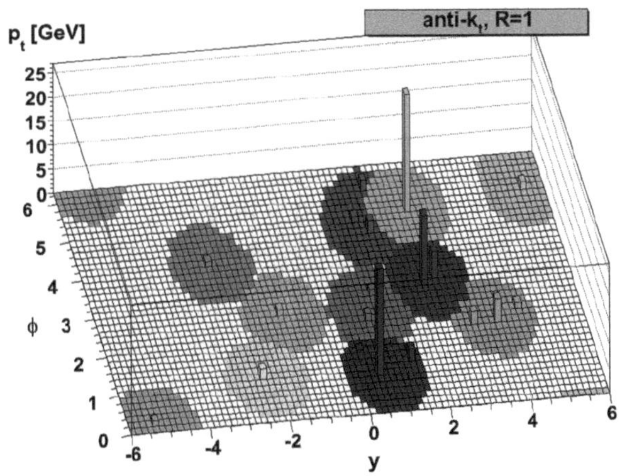

Fig. 5.6 The result of applying the anti-k_T algorithm to a generated parton level event with additional soft radiation. The *colored areas* indicate the covered areas of each jet. Harder jets take up a larger portion of the area than nearby softer jets [13]

pile-up and underlying event per unit area is ρ, the corrected four-vector of the jet is defined as:

$$p_c^\mu = p^\mu - A^\mu \rho, \tag{5.7}$$

where p^μ is the original four vector component and A^μ is the generalized "four-vector" jet area. The main ingredient in applying the correction is estimation of the UE+PU density ρ for each event. In the presence of PU interactions, there will be many jets that will be very soft. In the limit that the PU+UE contribution is dense and uniform, any PU jet will have a ratio of p_T divided by area equal to ρ. Since in principle PU/UE jets cannot be separated from hard jets in the events, ρ can be defined as:

$$\rho = \mathrm{median}\left[\frac{p_\mathrm{T}}{A}\right], \tag{5.8}$$

where the median is calculated for each jet in the event. In general, PU jets are more numerous than hard jets giving the average PU/UE contribution ρ that can be used to correct the hard jets in the event.

Additional jet energy corrections are used to correct the non-uniformity and non-linear response of the detector. Although particles are already corrected in PF level, additional corrections take into account any loss of particles in the reconstruction phase as a function of jet transverse momenta and direction. Several corrections are applied to account for the effects of pileup (L1) and make the jet response uniform in the detector (L2) and in different ranges of jet transverse momenta (L3). The corrections are factorized as:

$$E_c = C_{\mathrm{L1}} \times C_{\mathrm{L2\ Relative}}(\eta, p'_\mathrm{T}) \times C_{\mathrm{L3\ Absolute}}(p''_\mathrm{T}). \tag{5.9}$$

The L1 Fastjet correction is denoted as C_{L1}, while $C_{\mathrm{L2\ Relative}}(\eta, p'_\mathrm{T})$ and $C_{\mathrm{L3\ Absolute}}(p''_\mathrm{T})$ provide the L2 Relative and L3 Absolute corrections respectively. Each correction is applied to the four vector produced by the previous correction. Therefore, p'_T corresponds to the transverse momentum as estimated by L1 Fastjet and p''_T corresponds to the transverse momentum after L1 and L2. The L2 Relative correction is created to equalize the jet response as a function of η and remove effects from varying detector response along the detector. L2 is designed so that the response of a jet in an arbitrary η region is equalized to the response of a central jet. L2 corrections are obtained using data driven di-jet balance techniques. The L3 Absolute corrections equalize the response as a function of p_T so that the transverse momentum of the jet is corrected back to the particle level. This correction is obtained using $\gamma+$ jet or Z+jet events to correct the p_T of the jet to the well measured p_T of the photon or the Z respectively.

In the case of data, one additional jet correction, the L2L3 Residual correction is introduced. While the jet response in the simulation is very accurate, this small residual correction is applied to correct for the difference between data and simulation as a function of p_T and η. The L2L3 Residual correction is applied on top of all the other corrections.

5.6.3 Missing E_T Reconstruction

Missing transverse energy is used to identify the presence of invisible particles in the detector, mainly neutrinos, but also search for exotic particles appearing in processes beyond the SM (i.e. SUSY). Missing transverse energy is defined using PF candidates as:

$$\vec{\not{E}}_T = -\sum_{\text{particle}} \vec{p_T}, \tag{5.10}$$

where the sum runs over the transverse momentum vector (p_x, p_y) of all reconstructed PF candidates.

5.7 Identification of Jets Originating by *b* Quarks

Jets that arise from bottom quark hadronization and decay (b-jets) are present in a wide range of interesting physics processes such as the decay of top quark, Higgs bosons and various SUSY cascades. In this analysis, b-jets are extensively used to enhance the sensitivity for MSSM Higgs bosons that are produced in association with b-quarks. Properties of the bottom, and to lesser extent, charm quarks, such as their fragmentation functions and the relatively large mass and long lifetime of the heavy hadrons are used to identify the hadron jets into which they fragment. In addition we exploit the semileptonic decays of the heavy quarks by identifying muons and electrons inside those jets.

The CMS b-jet identification starts from reconstructed PF jets clustered with the anti-k_T algorithm. High quality tracks are selected inside the jet by requiring a transverse momentum of at least 1 GeV, at least eight hits associated to the track and a good track fit, with $\chi^2/\text{ndof} < 5$. The track parameters are extracted from all PF charged particles inside the jet. Since track measurements in the vicinity of the interaction vertex contain most of the discriminating power, at least two hits are required in the pixel system. A loose selection is applied on the track impact parameters to further increase the purity and reduce contamination by decay products of long lived particles like K_0s. The transverse and longitudinal impact parameters, d_{xy} and d_z are defined as the transverse and longitudinal distance to the primary vertex at the point of closest approach to the beam line. Their norms are required to be smaller than 0.2 and 17 cm respectively. Tracks are associated to jets in a cone $\Delta R < 0.5$ around the jet axis and at the point of closest approach, the distance to the jet axis is required to be smaller than 700 μm and this point is required to be within 5 cm from the primary vertex. Several algorithms are implemented in CMS for b-jet identification [15]. This analysis is using the Track Counting Algorithm.

5.7.1 The Track Counting Algorithm

The track counting (TC) algorithm exploits the impact parameters of the associated tracks to the jet. The impact parameter (IP) is calculated in three dimensions, taking advantage of the excellent resolution of the pixel detector along the z axis. A sign is assigned to the impact parameter based on the scalar product of the vector pointing from the primary vertex to the point of closest approach with the jet direction: tracks originating from the decay of particles that are traveling in the jet direction tend to be assigned to positive values while prompt tracks have positive or negative signs. The resolution of the impact parameter depends strongly on p_T and η of the track. Therefore, for discrimination purposes, the impact parameter significance is defined as $IP_{\text{sig}} = IP/\delta IP$, where δIP is the IP measurement error. Figure 5.7 shows the impact parameter and its significance for the selected tracks. The impact parameter

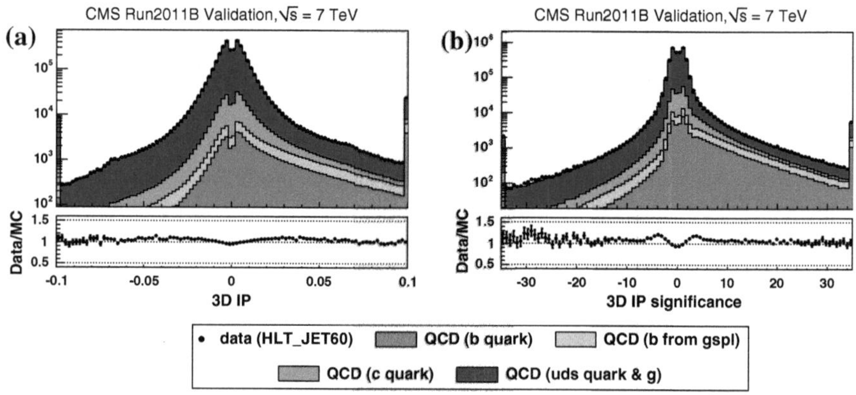

Fig. 5.7 Value (**a**) and significance (**b**) of the impact parameter for all selected tracks [15]

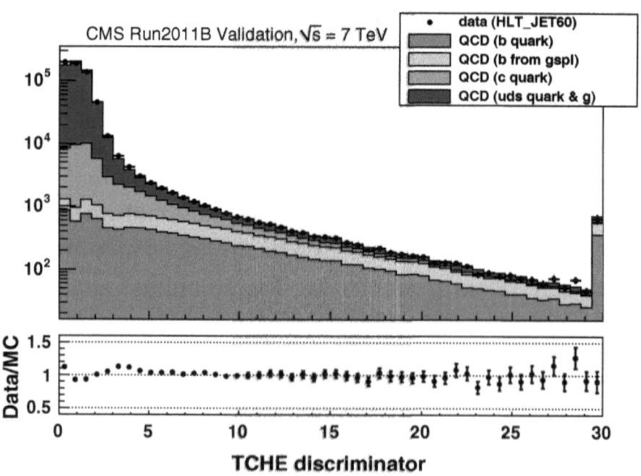

Fig. 5.8 Discriminator values for the TCHE algorithm [15]

significance alone has discriminating power between the decay projects of b and non-b jets. The *Track Counting* (TC) algorithm sorts tracks in a jet by decreasing values of the IP significance. While the ranking tends to bias the values for the first track to high positive IP significances, the probability to have several tracks with high positive values is low for light jets. This algorithm uses the IP significance of the second ranked track as the discriminator value and is referred as Track Counting High Efficiency (TCHE) algorithm. The values of the discriminator are presented in Fig. 5.8. Very good agreement is observed between data and simulation for the 2011 dataset corresponding to the data used in this analysis.

References

1. The CMS Collaboration, Tracking and primary vertex results in first 7 TeV collisions. CMS Physics Analysis Summary, CMS-PAS-TRK-10-005 (2010)
2. R. Fruhwirth, Application of kalman filtering to track and vertex fitting. Nucl. Instrum. Meth. **A262**(2), 444 (1987)
3. W.Erdmann, Offline primary vertex reconstruction with deterministic annealing clustering. CMS-IN-11-014, IN 2011–014 (2011)
4. K. Rose, Deterministic annealing, clustering, and optimization. Ph.D. Dissertation, California Institute of Technology, 1991
5. S. Baffioni, C. Charlot, F. Ferri, D. Futyan, P. Meridiani, I. Puljak, C. Rovelli, R. Salerno, Y. Sirois, Electron reconstruction in CMS. Eur. Phys. J. C: Part. Fields **49**(4), 1099–1116 (2007). doi:10.1140/epjc/s10052-006-0175-5
6. CMS Collaboration, Particle-flow event reconstruction in CMS and performance for Jets, Taus, and MET. CMS Physics Analysis Summary, CMS-PAS-PFT-09-001 (2009)
7. CMS Collaborationm, Commissioning of the particle-flow event reconstruction with leptons from J/Psi and W decays at 7 TeV. CMS Physics Analysis Summary, CMS-PAS-PFT-10-003 (2010)
8. W. Adam, R. Frühwirth, A. Strandlie, T. Todorov, Reconstruction of electrons with the Gaussian-sum filter in the CMS tracker at the LHC. J. Phys. G Nucl. Part. Phys. **31**(9), N9 (2005)
9. The CMS Collaboration, Studies of tracker material. CMS Physics Analysis Summary, CMS-PAS-TRK-10-003 (2010)
10. The CMS Collaboration, Performance of CMS Muon reconstruction in PP collisions at $\sqrt{s} =$ 7 TeV. CMS Physics Analysis Summary, CMS-PAS-MUO-10-004 (2011)
11. The CMS Collaboration, Muon high level trigger in CMS. CMS Physics Analysis Summary, CMS-PAS-MUO-10-003 (2010)
12. CMS Collaboration, Commissioning of the Particle-flow reconstruction in Minimum-Bias and jet events from PP collisions at 7 TeV. CMS Physics Analysis Summary, CMS-PAS-PFT-10-002 (2010)
13. Matteo Cacciari, Gavin P. Salam, Gregory Soyez, The anti-k_t jet clustering algorithm. J. High Energy Phys. **2008**(04), 063 (2008)
14. Matteo Cacciari, G.P. Salam, Pileup subtraction using jet areas. Phys. Lett. **B659**, 119–126 (2008)
15. The CMS Collaboration, B-jet identification in the cms experiment. CMS Physics Analysis Summary, CMS-PAS-BTV-11-004 (2011)

Chapter 6
Hadronic Tau Identification and Trigger

High performance hadronic tau identification is crucial for measurements and searches of new physics in final states involving tau leptons. In about two thirds of the cases, taus decay hadronically, typically into either one or three charged mesons (predominantly π^+, π^-) in presence of up to two neutral pions, decaying via $\pi^0 \rightarrow \gamma\gamma$. A similar signature is occasionally observed for quark and gluon jets. Due to the huge jet-production cross section, the experimental challenge in reconstructing and identifying genuine hadronic tau decays is formidable. This chapter describes the Hadron Plus Strips (HPS) Tau identification algorithm that was designed specifically for $Z/H \rightarrow \tau\tau$ analysis presented in the following chapters as well as the trigger algorithm optimized for di-tau final states.

6.1 Decay Mode Based Reconstruction

The traditional tau identification algorithms start from a jet and apply narrowness and isolation requirements on the jet to be tagged as a hadronically decaying tau. Usually a set of cones is used, a signal cone around the core of a jet or the leading track inside a jet, and a concentric isolation cone that defines an annulus that is used for isolation. One problem of the cone approach is the definition of the tau energy. Since the signal cone is usually designed to envelope the tau constituents, more particles from the underlying event are usually added, resulting in overestimation of the tau energy. This overestimation becomes much more significant in the case of high luminosity where pile-up is significant and constituents coming from other vertices could be associated with the tau. A second problem is the definition of the isolation. Since the background just comes from jet production, depending on how the jet hadronization fluctuates, it is possible that a jet is very narrow and falls into the signal cone. If additional constituents are associated to the tau lepton, they will not be used in calculating the isolation that is applied outside the cone. Both those problems can be solved by utilizing reconstruction of the actual decay modes. The

M. Bachtis, *Heavy Neutral Particle Decays to Tau Pairs*,
Springer Theses, DOI: 10.1007/978-3-319-03257-3_6,

tau decays via $\rho(770)$ or $a_1(1260)$ resonances that can be individually reconstructed inside the jet. The isolation sums can be created by using all particles that are not associated with the decay mode. This approach gives optimal separation of the signal and isolation constituents, providing simultaneously optimal definition of the tau energy and maximum rejection of background. Figure 6.1 illustrates the differences between the decay mode and the cone based approach for a jet that has a signature similar to a $\rho \rightarrow \pi^+\pi^0$ plus some additional particles. In the illustrated example, the fake tau candidate appears less isolated in the decay mode approach since no isolation annulus is required.

6.2 The HPS Tau Identification Algorithm

The Hadron plus Strips (HPS) tau identification algorithm is based on charged hadrons and neutral electromagnetic objects ("photons") that are provided by the CMS PF reconstruction. The HPS algorithm starts from a PF jet and searches for tau lepton decay products produced by any of the hadronic decay modes enumerated in Table 1.1. As neutral pions are produced very often in hadronic tau decays, one focus of the HPS tau identification algorithm is an optimized π^0 reconstruction. Special attention is paid to photon conversions in the CMS tracker material, which typically result in broader calorimeter signatures for neutral pions in the azimuthal direction, due to the bending of electron/positron tracks in the 3.8 T solenoidal magnetic field of CMS.

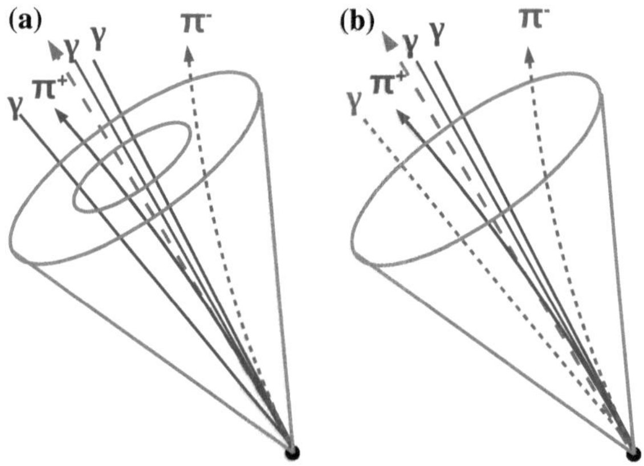

Fig. 6.1 Illustration of the different approach of cone based (**a**) and decay mode based (**b**) tau identification in a jet that has a $\rho \rightarrow \pi^+\pi^0$ candidate signature and additional particles. Constituents that are shown as *solid lines* contribute to the signal cone for the tau, while constituents shown as *dotted lines* correspond to candidates that contribute to the isolation

6.2.1 Reconstruction of π^0 Sub-Clusters

The possible broadening of calorimeter signatures by photon conversions is accounted for in the HPS algorithm by reconstructing photons in "strips", objects which are built out of electromagnetic particles. The strip reconstruction starts by centering one strip on the most energetic electromagnetic particle reconstructed by the PF algorithm within the jet. The algorithm then searches for other electromagnetic particles within the window of size $\Delta\eta = 0.05$, $\Delta\phi = 0.20$ around the strip center. If other electromagnetic particles are found within that window, the most energetic of these particles is associated with the strip, and the strip center position is recalculated to match the sum of four-vectors of all PF particles associated with the strip. The next highest energetic electromagnetic particle is then searched for within the $\Delta\eta = 0.05$, $\Delta\phi = 0.20$ window around the new strip center until no further particles can be associated with the strip. In this case the algorithm creates a new strip centered around the most energetic particle not associated with any strip. Strips satisfying a minimum transverse momentum requirement of $p_{\mathrm{T}}^{strip} > 1\,\mathrm{GeV/c}$ are finally combined with the charged hadrons to reconstruct individual hadronic tau lepton decay modes.

6.2.2 Combination of Charged Hadrons and Strips

The decay modes which are considered by the HPS tau identification algorithm are:

1. Single hadron: This signature reconstructs $\tau^- \rightarrow h^-\nu_\tau$ decays and $\tau^- \rightarrow h^-\pi^0\nu_\tau$ decays in which the neutral pions have too little energy to be reconstructed as strips.
2. One hadron + one strip: This signature aims to reconstruct the decay mode $\tau^- \rightarrow h^-\pi^0\nu_\tau$ in events in which the impact position of the photons from π^0 decays are close together on the calorimeter surface. The size of the strip in ϕ direction is large enough to account for the possibility that one or both of the photons have converted.
3. One hadron + two strips: This signature aims to reconstruct the decay mode $\tau^- \rightarrow h^-\pi^0\nu_\tau$ in events in which the impact positions of photons from π^0 decays are well separated on the calorimeter surface.
4. Three hadrons: This signature aims to reconstruct the decay mode $\tau^- \rightarrow h^-h^+h^-\nu_\tau$. The three charged hadrons are required to have charge sum $|q| = 1$ and to be refitted to the same secondary vertex, reconstructed by the Kalman vertex fit algorithm.

The four-vectors of hadrons and strips are reconstructed according to the respective tau decay mode hypothesis and are required to be compatible with the masses of intermediate meson resonances listed in Table 1.1. In case of the one hadron + one strip signature the four-vector reconstructed from the strip energy and position

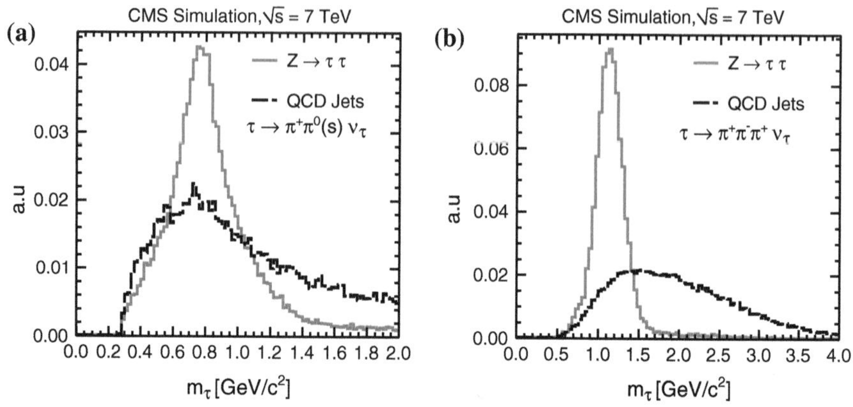

Fig. 6.2 Reconstructed visible tau mass for tau decays. **a** one prong + strip **b** three prongs

is set to match the nominal π^0 mass. Whereas in the case of the hadron + two strips signature, a massless four-vector is reconstructed for each strip and the invariant mass of the two strips is required to be within 50 and 200 MeV/c^2. The signature of three charged hadrons is not affected by photon conversions and is well identified. Figure 6.2 shows the reconstructed visible mass of the hadronic tau candidates for simulated taus produced in $Z \rightarrow \tau\tau$ and in fake taus reconstructed in QCD jets. In the cases of hadrons+ strips and three prong decays, the vector meson mass reconstructed with the one prong + strip decay mode is broader than the three prong decays due to photon conversions and energy loss of the π^0. In the case of one prong +1 strip the invariant mass of the tau is required to be between 0.3 and 1.3 GeV/c. In the case of a hadron +2 strips, the invariant mass is required to be within 0.4 and 1.2 GeV/c. In the case of three prongs, where only track reconstruction is performed, the a_1 signature is very clean and an invariant mass window between 0.8 and 1.5 GeV/c is required.

For additional background rejection, narrowness criteria are required. Figure 6.3 shows the cone size in terms of ΔR or opening angle as a function of the reconstructed tau energy or transverse momentum for real taus from $Z \rightarrow \tau\tau$ decays. A narrowness criterion is defined by requiring that the cone of the hadronic tau is smaller than $\Delta R = 2.8/p_{\mathrm{T}}^{\tau}$, where p_{T}^{τ} is computed by summing the four-vectors of reconstructed charged hadrons plus strips, where the cone is defined by the maximum distance of the associated constituents and the tau four-vector. In addition the cone is not allowed to exceed $\Delta R = 0.1$. To account for the material effects the minimum restricting cone cannot be smaller than $\Delta R = 0.05$. In addition, the hypothetical visible tau momentum p^{τ} is required to match the (η, ϕ) direction of the seeding jet within a maximum distance of $\Delta R = 0.1$. If more than one hypothesis for tau decay signatures exists after all selection requirements, the hypothesis leading to the lowest E_T sum of jet constituents not associated with tau decay products is given preference

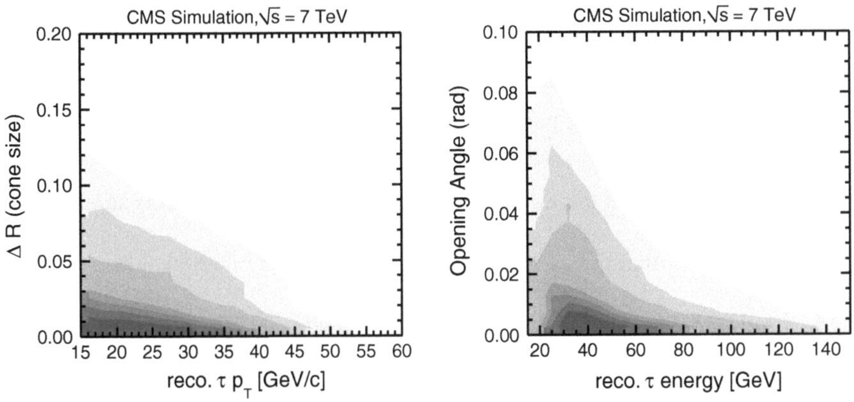

Fig. 6.3 Tau decay cone size (ΔR) versus reconstructed tau P_{T} (*left*) and tau opening angle (*rad*) versus reconstructed tau energy (*right*)

(most isolated candidate) to ensure that there is only one tau reconstructed per PF Jet.

6.2.3 Hadronic Tau Isolation

Tau isolation is the major discriminator against QCD jets. Isolation is defined in a cone of $\Delta R = 0.5$ around the tau four-vector. All PF charged and neutral electromagnetic candidates in the cone contribute to the isolation sums except for the reconstructed decay mode constituents. The total isolation transverse energy is defined as:

$$I_{\tau}^{PF} = \sum P_{\mathrm{T}}^{\text{charged}} + \sum E_{\mathrm{T}}^{\gamma} \tag{6.1}$$

where the sums are running on the non associated PF charged and PF γ candidates. Due to the higher instantaneous luminosity resulting in a large number of pile-up interactions observed in the detector, isolation is designed to be robust against pile-up by introducing the $\Delta\beta$ corrections in Tau ID, as implemented for muons and electrons in Sect. 5.5. In the case of taus, neutral hadrons are not used. Therefore, the correction factor is computed such that the efficiency is almost flat as a function of the number of reconstructed vertices. The corrected $\Delta\beta$ isolation sum is defined as

$$I_{\tau}^{PF/\Delta\beta} = \sum P_{\mathrm{T}}^{\text{charged}} + \max\left(\sum E_{\mathrm{T}}^{\gamma} - f^{\Delta\beta} \sum E_{\mathrm{T}}^{\text{charged}}(\mathrm{PU}), 0.0\right) \tag{6.2}$$

where $\sum_{\text{PFcharged}} E_T(\mathrm{PU})$ is the sum of the charged particle transverse energy coming from pile-up interactions in a cone of $\Delta R = 0.8$ around the tau four vector and $f^{\Delta\beta}$ is the $\Delta\beta$ correction factor. The PU contribution is summed in a cone of 0.8 so that there are more statistics available for the correction. The $\Delta\beta$ factor is selected to account for the effect of the larger PU cone with respect to the nominal isolation cone and was chosen to be $f^{\Delta\beta} = 0.458$.

Several isolation working points are defined by imposing different requirements on $I_T^{\text{sum}}(\Delta\beta)$:

- **VLoose**: Requires that the $\Delta\beta$ corrected isolation sum($I_T^{\text{sum}}(\Delta\beta)$ is smaller than 3.0 GeV.
- **Loose**: Requires that the $\Delta\beta$ corrected isolation sum($I_T^{\text{sum}}(\Delta\beta)$ is smaller than 2.0 GeV.
- **Medium**: Requires that the $\Delta\beta$ corrected isolation sum($I_T^{\text{sum}}(\Delta\beta)$ is smaller than 1.0 GeV.
- **Tight**: Requires that the $\Delta\beta$ corrected isolation sum($I_T^{\text{sum}}(\Delta\beta)$ is smaller than 0.8 GeV.

The performance of those working points is presented in the following sections.

6.3 Discrimination Against Electrons and Muons

Although the most significant backgrounds to analyses with taus come from jets, real electrons and muons also contribute significantly to the background, especially in the case of the di-tau analysis where direct Drell-Yan production of di-muons or di-electrons fakes a di-tau signature when one tau decays to electron or muon and the other hadronically. To reject this type of background, specialized electron and muon rejection criteria are implemented.

6.3.1 Muon Rejection

Muons fake taus by being reconstructed as single prong taus consisting of a perfect track. Charged pions can be discriminated against muons by checking the number of muon segments associated with the leading track of the tau. Due to the very high efficiency of the muon detectors, segment requirements reject almost all real muons. However, there are cases where the muon has passed through cracks in the muon system therefore leaving no segments in the muon chambers. To reject the remaining muons, the compatibility between the calorimeter energy and the track momentum is required to be consistent with the pion hypothesis. Two operating points are defined for muon rejection:

- **Loose Muon Rejection**: Requires that the leading track of the tau is not associated with a global muon or a tracker muon with at least one segment match.
- **Tight Muon Rejection**: On top of the loose working point, checks for calorimeter compatibility the single prong decay model. The ECAL associated energy ($E_{\text{ECAL}}^{\text{PF}}$)

and HCAL associated energy ($E^{\mathrm{PF}}_{\mathrm{HCAL}}$) from the Particle Flow algorithm is used to define the compatibility. The leading charged candidate of the tau is required to have $\left(E^{\mathrm{PF}}_{\mathrm{ECAL}} + E^{\mathrm{PF}}_{\mathrm{HCAL}}\right) / P_{\mathrm{track}} > 0.1$ where P_{track} is the track momentum.

6.3.2 Electron Rejection

Separation between electrons and taus is very challenging due to the fact that a showering electron has a very similar signature with a hadronically decaying tau. The Particle Flow algorithm has intrinsic electron/pion discrimination ($\xi(e/\pi)$) implemented for identifying electrons inside jets. This discriminator has a pion mis-identification rate of about 15 %. The huge background contamination from $Z/\gamma^* \rightarrow ee$ in the electron + tau final state defines the need for additional discrimination against electrons. Electrons that are faking taus can be identified by the fraction of their electromagnetic energy. Therefore the performance and coverage of the electromagnetic calorimeter is very important. Electrons that fall in the ECAL crack between barrel and endcap and in the boundaries of the ECAL supermodules leave no energy in the ECAL making discrimination very difficult. Because the cracks contribute to a very small percentage of the total coverage, taus with compatible directions with the cracks are rejected. For additional electron rejection a multivariate discriminator is used and is described in detail elsewhere [1]. The multivariate discriminator exploits the electromagnetic fraction of the tau, the PF electron/pion discriminator, the shower profile of the electromagnetic energy and the properties and presence of a GSF track. The electron rejection working points are defined as follows:

- **Loose**: Requires that the leading charged candidate of the tau has a PF electron compatibility of $\xi(e/\pi) < 0.6$
- **Medium**: Requires that the leading charged candidate of the tau is not identified as a PF electron $\xi(e/\pi) < -1$ and that the tau direction is incompatible with the ECAL crack and the super module boundaries
- **MVA**: Requires that the multivariate discriminator has passed and that the tau direction is incompatible with the ECAL crack and supermodule boundaries.

6.4 Performance of Tau Identification

Due to the fact that the main background to taus comes from jets that are produced in processes with many orders of magnitude higher cross section, high performance tau identification is crucial to make possible the analysis with tau leptons. The main measures of performance is the efficiency and the fake rate. The tau identification efficiency is defined as:

$$\epsilon_{\tau} = \frac{N_{\tau}(\mathrm{ID})}{N_{\tau}(\mathrm{ref})}, \tag{6.3}$$

where $N_{\tau}(\text{ref})$ is the number of simulated real tau objects and $N_{\tau}(\text{ID})$ is the number of simulated real tau objects that have passed the HPS Tau identification requirements. In the case of simulation, the true tau object is defined as a simulated tau lepton that has decayed hadronically and its four-vector is defined by the visible products of the tau decay excluding the neutrinos.

The tau fake rate quantifies the background and is defined per jet as:

$$f_{\tau} = \frac{N_{\text{jet}}(\text{ID})}{N_{\text{jet}}(\text{ref})}, \tag{6.4}$$

where $N_{\text{jet}}(\text{ref})$ is the number of PF jets reconstructed with the anti-kt algorithm (parameter 0.5) in the acceptance of the CMS tracker ($|\eta| < 2.5$) and $N_{\text{jet}}(\text{ID})$ is the number of reference jets that have also been identified as taus. The fake rate is measured using data and is compared to the expectation from MC simulation. The efficiency is defined using simulated $Z \rightarrow \tau\tau$ events in this section and a measurement using data is described in Chap. 8 with the scope of the $Z \rightarrow \tau\tau$ cross section measurement.

6.4.1 Tau Performance in Simulation

The tau identification efficiency is defined with respect to simulated taus produced in $Z \rightarrow \tau\tau$ decays. Generated taus that decay hadronically, having a visible transverse momentum of $p_T > 5$ GeV/c and $\eta < 2.3$ are selected. Reconstructed taus are matched to generated objects using a cone of $\Delta R = 0.1$. Figure 6.4 shows the tau reconstruction efficiency as a function of the generated p_T and η for the decay mode reconstruction and all isolation working points for reconstructed tau transverse

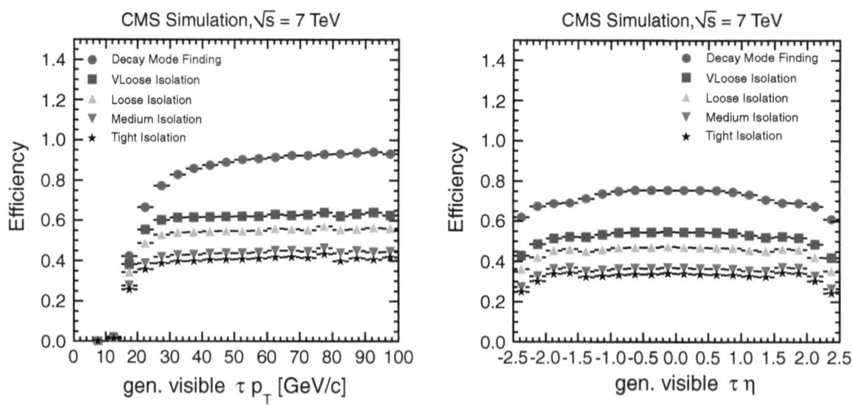

Fig. 6.4 Tau Identification efficiency as a function of generated visible tau p_T (*left*) and η (*right*) for taus produced in Z decays using POWHEG simulation interfaced with TAUOLA for modeling the tau decays

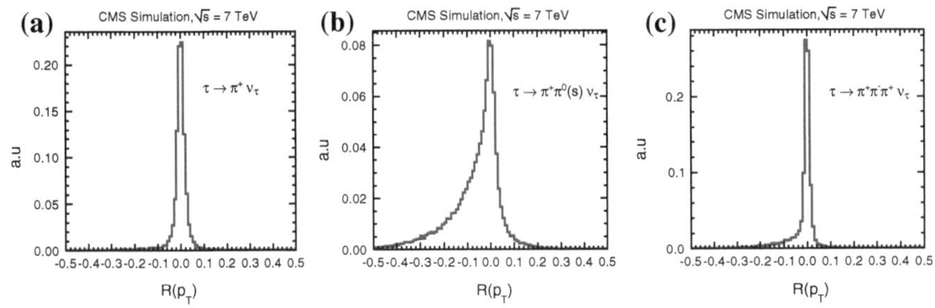

Fig. 6.5 Tau transverse momentum resolution ($R(p_\mathrm{T})$) for different tau decay modes. **a** one prong **b** one prong + strips **c** three prong

momentum of $p_T > 20\,\mathrm{GeV/c}$. The HPS reconstruction shows very steep turn-on behavior, establishing an efficiency plateau of about 50 % for the Loose isolation working point used in this analysis. In addition, the efficiency is flat as a function of the pseudorapidity for most part of the detector, falling at the most forward region due to tracking inefficiency and material effects in π^0 reconstruction.

Another additional performance metric is the reconstruction of the tau four vector quantified by the transverse momentum resolution defined as

$$R(p_\mathrm{T}) = \frac{p_\mathrm{T}^\mathrm{reco} - p_\mathrm{T}^\mathrm{gen}}{p_\mathrm{T}^\mathrm{gen}}, \tag{6.5}$$

where $p_\mathrm{T}^\mathrm{reco}$ is the reconstructed tau decay mode p_T and $p_\mathrm{T}^\mathrm{gen}$ is the generated visible tau p_T. Figure 6.5 shows the transverse momentum resolution for each individual decay mode of the HPS algorithm. In the case of one prong and no strips, the tau is a single charged hadron. As a result, the measurement is dominated by the excellent P_T resolution of the CMS tracker. In the case of a hadron + strips there is an energy loss tail resulting from the challenge in reconstructing low energy photons in the presence of material. Finally, in the three prong case the reconstruction is dominated by track triplet reconstruction which is improved by vertex refitting. The precision in the energy measurement using the decay mode approach shows great improvement compared to previous proposed reconstruction methods [2].

6.4.2 Measurement of the Tau Fake Rate

The tau fake rate is measured in two samples relevant for this analysis, muon enriched QCD and W+jets. The muon enriched sample requires a reconstructed tight muon with $p_T > 25\,\mathrm{GeV/c}$ and $\eta < 2.1$. The muon is required to not be isolated having $I_\mathrm{PF}^\mathrm{rel} > 0.5$ with isolation as defined in Sect. 5.5. To reject contamination from Drell-Yan muon pairs, a veto is applied on a second muon with $p_T > 15\,\mathrm{GeV/c}$ in

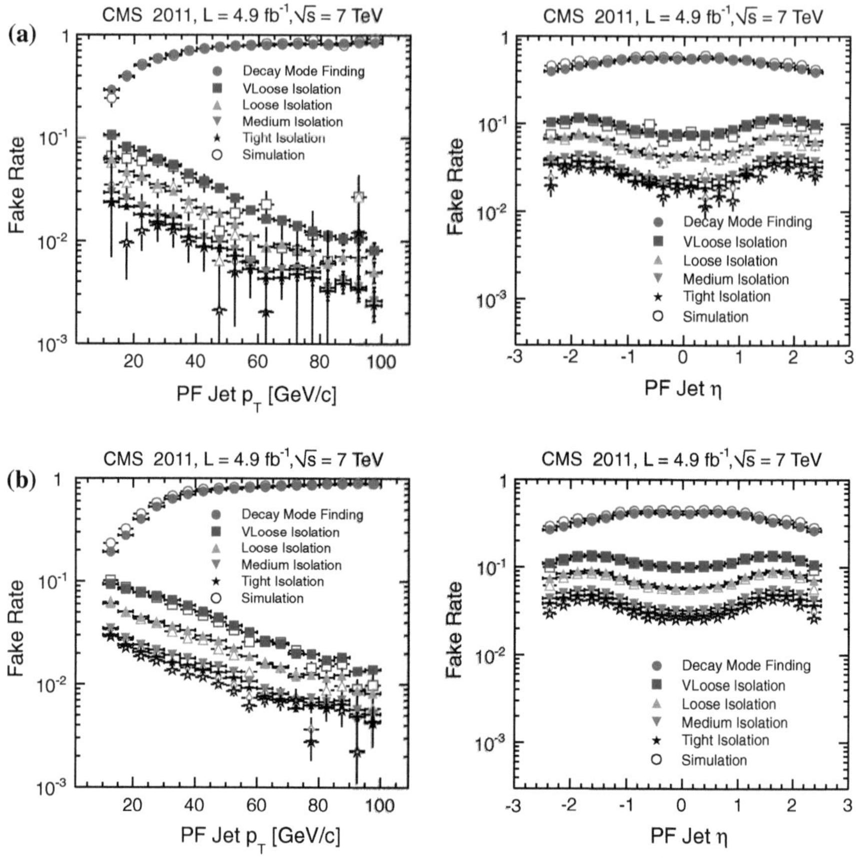

Fig. 6.6 Tau fake rate as a function of PF jet p_T (*left*) and η (*right*) for a muon enriched QCD sample and a W+jets enriched sample bottom. Filled points corresponds to CMS 2011 data and hollow points to PYTHIA simulation **a** muon enriched **b** W + jets

the event. The Tau ID is tested against jets that don't overlap with the muon by a distance of $\Delta R > 0.5$. The second major background to $Z/H \rightarrow \tau\tau$ analysis is W production in association with jets, which fake a tau. A W sample is selected by requiring a tight muon with $p_T > 25$ GeV/c and $\eta < 2.1$ that is required to be isolated requiring $I_{\mathrm{PF}}^{\mathrm{rel}} < 0.15$. The W selection is enhanced by requiring that $M_T(\mu, PF\ \not{E}T) > 65$ GeV, where $M_T(\mu, PF\ \not{E}_T)$ is the transverse mass of the muon and the missing transverse energy defined as

$$M_T(\mu, \not{E}_T) = \sqrt{2P_{\mathrm{T}}^{l}\ \not{E}_T \cdot (1 - \cos\Delta\phi)}. \quad (6.6)$$

For rejection of Drell Yan background, a veto is applied in the presence of a second muon with $p_T > 15$ GeV/c in the event.

Figure 6.6 shows the fake rate as a function of the tested jet p_T and η for both muon enriched and W+jets samples. The Loose Isolation working point (Sect. 6.2.3 provides a fake rate that is smaller than 1 % for both samples). The fake rate shows increase in the region near the barrel—endcap boundary for the isolation working points where there is large amount of material. In this region, the $\Delta\beta$ corrections are higher because photons from π^0 have higher conversion rates, resulting in the charged particle sums from PU vertices being higher implying larger corrections.

6.5 Trigger for di-tau Final States

Due to the high luminosity achieved by the LHC in 2011 run, a dedicated trigger was introduced for final states with a lepton and a hadronically decaying tau. The main purpose of this trigger was to maintain very high efficiency for the analysis without biasing offline tau identification. The main challenge in achieving this goal was the use of tau algorithms at L1. The L1 trigger uses only calorimeter information. Therefore a measurement of the tau with the calorimeter would bias the precise measurement of the decay mode reconstruction that was described in previous sections. On the other hand, the presence of a light lepton (electron or muon) in the final state and the better performance of the L1 trigger for electrons and muons makes possible the use of single electron or muon trigger at the L1 with the additional requirement of the tau trigger only in the High Level Trigger.

6.5.1 High Level Trigger Algorithm

All the events that pass single muon or electron L1 trigger are passed as input to the HLT. The next step is full online reconstruction of the electron or muon in the HLT. Muons and electrons are reconstructed with similar algorithms as the full offline breconstruction and several identification and isolation criteria are applied to reject fake objects and reduce the rate. Only the events that pass the single electron or muon trigger are forwarded to the Tau HLT. At this point the rate has been reduced significantly which makes possible the use of sophisticated algorithms for tau identification.

The first Tau HLT step is identification of points of interest for hadronic taus by clustering jets using only information from ECAL and HCAL. A loose requirement of $E_T > 5$ GeV is applied on the jets so that the energy measurement is not biased. The jets that are selected correspond to seeds for possible taus. The next step is pixel track and vertex reconstruction. Track reconstruction is performed in several steps during an iterative tracking process similar to the offline algorithm described in Sect. 5.1.2. First, tracks are reconstructed using pixel hits and vertices are formed by those tracks. If a pixel track has $P_{\mathrm{T}} > 0.9$ GeV it is used as seed for full track reconstruction. Alternatively, for pixel triplets above 0.5 GeV an association is formed between the seeding jets and the pixel tracks for tracks coming from the primary vertex and within

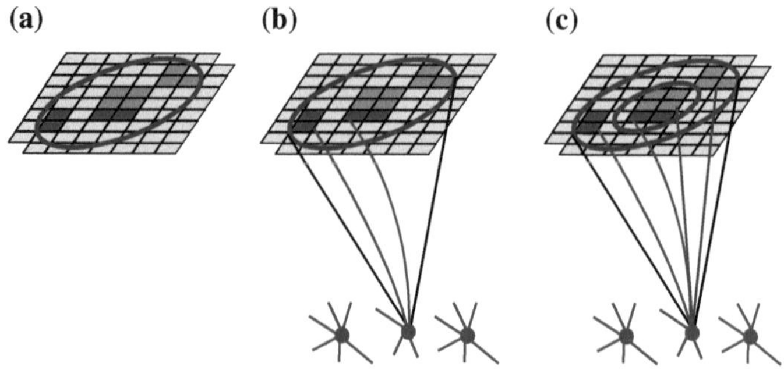

Fig. 6.7 Illustration of the different steps of tau identification in the HLT. **a** Calorimeter clustering, **b** Pixel vertex and track reconstruction, Jet-Track association and full silicon tracking. **c** Particle Flow reconstruction and tau ID. Bent lines correspond to charged hadrons and straight lines to PF photons

a window of $\eta \times \phi = 0.5 \times 0.5$. Tracks that satisfy the above requirements contribute to the next steps of the iterative tracking process. To maintain high efficiency in the presence of dead pixel channels, pixel tracks are formed by the vertex and a pixel pair. Those tracks with $p_T > 1.2$ GeV/c are used as seeds as well. The next iterations of the tracking use the existing tracks as seeds to provide better measurement and reject fake tracks. After tracking is performed, information from all relevant detectors is available to reconstruct hadronic taus. For maximum synchronization of the online and offline reconstruction, Particle Flow algorithm is run at this step using a lighter configuration. Calorimeter clustering is performed only in the regions of interest. Tracks and clusters are combined to build PF candidates in the same way as in the offline reconstruction. The PF candidates produced during online reconstruction are then used to form PF Jets. The PF Jets are then fed to the Tau identification algorithm to form tau candidates. A simpler and generic cone based algorithm is used to identify taus. A signal cone is formed around the leading PF Charged candidate that has a size of $\Delta R = 0.15$. All charged and neutral electromagnetic (PF photon) constituents in the signal cone are defining the tau four vector. Isolation is computed in an annulus around the leading charged hadronic candidate with a maximum cone size of $\Delta R = 0.5$. Many isolation working points are defined depending on the instantaneous luminosity and the final state:

- **Loose** Requires no charged candidates in the isolation annulus with $p_T > 1.5$ GeV/c.
- **Medium** Requires no charged candidates in the isolation annulus with $p_T > 1.0$ GeV/c.
- **Tight** Requires no charged candidates in the isolation annulus with $p_T > 1.0$ GeV/c and no neutral electromagnetic PF candidates in the annulus with $E_T > 1.5$ GeV.

An illustration of the different steps of Tau Identification in the HLT is presented in Fig. 6.7.

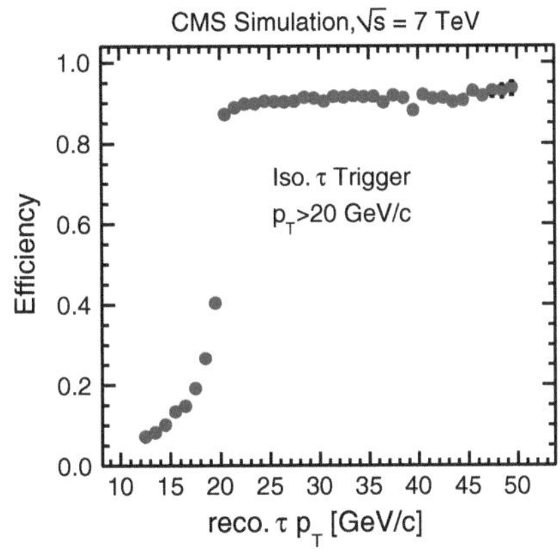

Fig. 6.8 Efficiency of the Isolated Tau trigger as a function of the reconstructed tau p_T where tau is produced in Z decays has been matched to the generator level

6.5.2 Performance

The performance of the Tau HLT algorithm is measured with respect to high quality reconstructed taus that pass offline tau identification. The offline tau is required to have been reconstructed using the HPS Algorithm within $\eta < 2.3$ and be isolated using the HPS Loose working point. In addition, the offline reconstructed tau is required to be matched to a generated tau object. The triggered tau is required to be above a threshold and be matched to the offline object within a distance of $\Delta R < 0.3$. The trigger efficiency is presented in Fig. 6.8 after requiring a trigger threshold of 20 GeV/c. The turn on curve is very steep due to the precise tau energy measurement made online and the efficiency reaches a flat value of about 90 %. The 10 % inefficiency observed is mainly due to mis-association of the track seeds and the primary pixel vertex and appears in the case when the offline and the pixel primary vertices are not the same. This inefficiency can be improved by defining the primary vertex as the vertex that is nearest to the triggered lepton in the lepton + tau trigger.

References

1. The CMS Collaboration, *Search for neutral Higgs bosons decaying to tau pairs in pp collisions at* $\sqrt{s} = 7$ TeV. (CMS Physics, Analysis Summary, 2011) (CMS-PAS-HIG-11-029)
2. The CMS Collaboration, Detector performance and software: physics technical design, report, 2006 (CERN-LHCC-2006-001)

Chapter 7
Selection of Tau Pairs

The first step of the analysis is the selection of high quality di-tau candidates that can be used for Z cross section measurement and Higgs searches. The analysis is performed in three final states which have at least one light lepton to facilitate easier trigger:

- $\mu + \tau_h$: Consists of one tau decaying to muon and the other decaying hadronically.
- $e + \tau_h$: Consists of one tau decaying to electron and the other decaying hadronically.
- $e + \mu$: Consists of one tau decaying to electron and the other decaying to muon.

The $\mu + \tau_h$ and $e + \tau_h$ final states have a branching ratio of about 23 %. The $\mu + \tau_h$ final state is relatively clean and the major background is heavy flavor QCD production and W+jets where a jet can fake a tau lepton. The $e + \tau_h$ final state is characterized by higher background contamination due to the increased fake rate of the electron compared to the muon. In addition since a hadronic tau decay via $\tau \rightarrow \rho\nu \rightarrow \pi^+\pi^0\nu$ can be dominated by the electromagnetic part, the tau signature in the detector is sometimes similar to the one of the electron, resulting in higher background contamination from $Z/\gamma^* \rightarrow ee$ compared to the $\mu + \tau_h$ final state. The $e + \mu$ final state consists of two opposite flavor leptons with a branching ratio of about 6 % but is almost background free. The small background in $e + \mu$ mainly includes $t\bar{t}$, heavy flavor QCD and W/Z+jets, where jets are faking leptons.

7.1 Data Samples and Triggers

Events are selected from the 2011 CMS data sample that corresponds roughly to a delivered LHC integrated luminosity of $\mathcal{L} = 5.2\,\text{fb}^{-1}$. Preselection is performed in the level of L1 Trigger. For the $\mu + \tau_h$ final state, a single muon trigger is required at L1 with a threshold that varies from 7 to 14 Gev depending on the instantaneous luminosity during the run. For the $e + \tau_h$ final state, L1 requires a single e/γ trigger with a threshold that varies from 12 to 20 Gev. Finally in the case of $e+\mu$ final state, a set of cross triggers is used utilizing asymmetric thresholds for the two electron legs.

M. Bachtis, *Heavy Neutral Particle Decays to Tau Pairs*,
Springer Theses, DOI: 10.1007/978-3-319-03257-3_7,

In the first set, an e/γ object of 12 Gev is combined with the bare presence of a L1 muon with no threshold (MuOpen) and the second set is using an e/γ object of 5 Gev combined with a muon with a threshold varying from 7 to 12 Gev. The events that are accepted by the requirements described above are passed to the High Level Trigger. The HLT, in the case of the $\mu+\tau_h$ final state, uses a combined muon+tau trigger with the muon threshold varying from 12 to 15 Gev and the tau threshold varying from 10 to 20 Gev. Loose isolation is applied on the tau trigger as described in Sect. 6.5. In the case of the $e+\tau_h$ final state, the rate is higher. Therefore an electron of 15–20 Gevis combined with a tau of 20 Gev with isolation varying between medium and tight as described in Sect. 6.5 . Finally, in the $e+\mu$ final state, two electron+muon cross triggers are applied depending on the L1 seeds requiring a leading lepton of 17 Gev combined with a sub-leading lepton of 8 Gev. The list of the HLT paths used and their respective L1 seeds are presented in Table 7.1.

After passing the HLT requirements, events are stored after they are categorized in primary datasets. The categorization of events into primary datasets is performed according to sets of trigger paths that are used for common analysis needs and also the cumulative rates of those paths. An event can be stored in multiple datasets if it fired different triggers. Since that implies that events can be stored twice, the primary datasets are tuned for minimal overlap. To collect the di-tau events in this analysis the TauPlusX and the MuEG datasets are used. The TauPlusX dataset includes cross triggers with a tau lepton while the MuEG dataset includes muon+electron cross triggers. For supporting studies of the muon and electron performance and background

Table 7.1 Summary of the used trigger paths for the $\mu+\tau_h$, $e+\tau_h$ and $e+\mu$ final states and their corresponding fractions of the dataset as described in the text

HLT Path	L1 Seed	Sample fraction (%)
$\mu\tau_h$-final state		
IsoMu12_LooseIsoPFTau10	L1_SingleMu7	0.5
IsoMu15_LooseIsoPFTau15	L1_SingleMu10	44.2
IsoMu15_eta2p1_LooseIsoPFTau20	L1_SingleMu14_eta2p1	55.3
$e\tau_h$-final state		
Ele15_LooseIsoPFTau20	L1_SingleEG12	25.0
Ele15_TightIsoPFTau20	L1_SingleEG12	15.9
Ele18_MediumIsoPFTau20	L1_SingleEG15	38.6
Ele20_MediumIsoPFTau20	L1_SingleEG18	20.4
Ele20_MediumIsoPFTau20	L1_SingleEG20	0.1
$e\mu$-final state		
Mu8_Ele17_CaloIdL	MuOpen_EG12	11.4
Mu8_Ele17_CaloIdT_CaloIsoVL	MuOpen_EG12	39.2
Mu17_Ele8_CaloIdL	Mu7_EG5	45.6
Mu17_Ele8_CaloIdT_CaloIsoVL	Mu12_EG5	3.8

normalization, the Single Muon, Single Electron, Double Muon and Double Electron datasets are also used (Table 7.1).

A data certification procedure ensures full and smooth detector operation and validation of the basic physics objects used in analysis. Several data taking periods that had improper detector operation are flagged as not certified and are not used for analysis. The total certified dataset that is used in this analysis corresponds to 4.934 fb^{-1} for the $\mu+\tau_h$ final state, 4.922 fb^{-1} for the $e+\tau_h$ final state and 4.979 fb^{-1} for the $e+\mu$ final state respectively.

7.2 Object Selection

The selected events are required to have at least one primary vertex with the requirements defined in Sect. 5.1.3. Muons are required to pass the tight requirements described in Sect. 5.3 and have transverse momentum above 17 Gev for the $\mu+\tau_h$ final state while, for the $e+\mu$ final state, asymmetric thresholds of 20 and 10 Gev are used depending on the cross trigger thresholds. Electrons are required to be reconstructed with the ECAL seeded method, pass the requirements described in Sect. 5.2, have $p_T > 20$ Gev for the $e+\tau_h$ final state and asymmetric thresholds of 20 and 10 Gev respectively, are used for the $e+\mu$ final state depending on which of the cross trigger fired. Muons and electrons are constrained within $\eta < 2.1$ for all final states. Both leptons are required to be isolated using $I_\ell^{PF/\Delta\beta} < 0.1$ (Sect. 5.5) for $e+\tau_h$ and $\mu+\tau_h$ final states, while in the case of the $e+\mu$ final state, a looser isolation requirement of $I_\ell^{PF/\Delta\beta} < 0.15$ is used. Hadronic taus are required to have $p_T > 20$ Gev and $\eta < 2.3$ and are identified using the methodology discussed in Chap. 6. Taus are required to pass Loose isolation discriminator using $\Delta\beta$ corrections. To reject Drell-Yan events where a light lepton fakes a hadronic tau decay, taus in the $\mu+\tau_h$ final state are required to pass tight discrimination against muon and loose discrimination against electrons. On the other hand, in the $e+\tau_h$ final state, taus are required to pass the MVA electron discriminator and loose muon discriminator. The transverse momentum spectra of muons, electrons and taus after basic selection are shown in Fig. 7.1–7.3 for the three considered final states.

7.3 Topological Requirements

After object selection, the main reducible backgrounds present in the analysis include QCD, W+jets and $t\bar{t}$. In the case of W and top pair production, the missing transverse energy is exploited to further reduce those backgrounds. In the case of the $\mu+\tau_h$ and $e+\tau_h$ final states, the transverse mass of the muon and missing transverse energy is formed as following:

$$M_T = \sqrt{2p_T^l \not{E}_T \cdot (1 - \cos\Delta\phi)}, \tag{7.1}$$

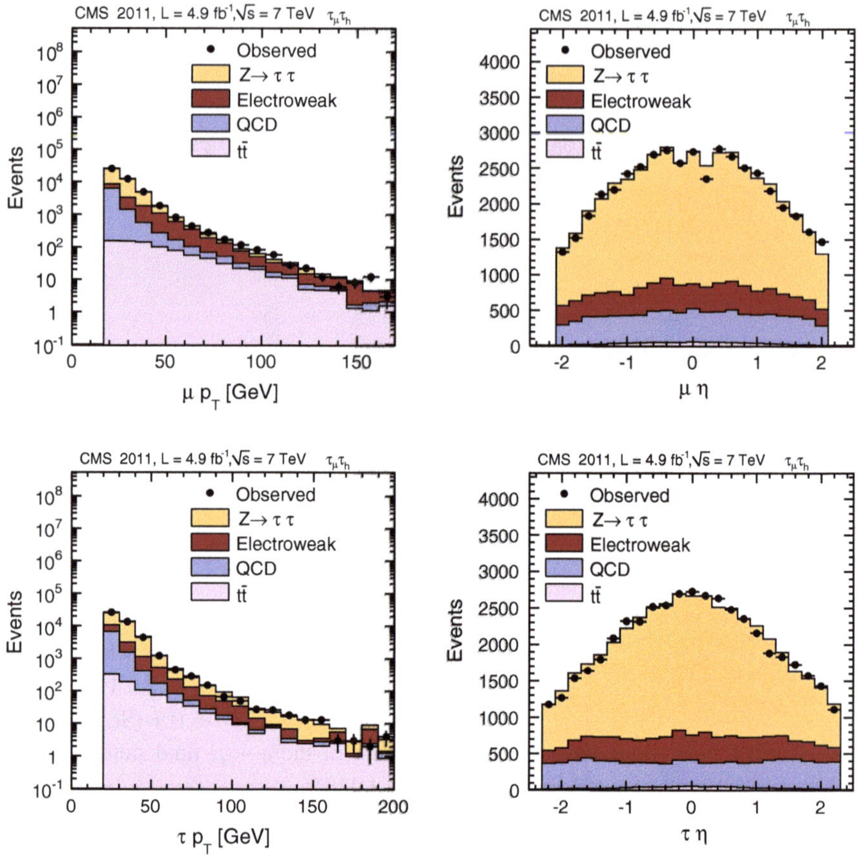

Fig. 7.1 Muon and Tau transverse momentum and pseudorapidity for the $\mu + \tau_h$ final state

where p_{T}^{l} is the lepton transverse momentum, $\not{E}_T$ is the PF missing transverse energy and $\Delta\phi$ is the acoplanarity angle between the missing energy transverse vector and the lepton transverse momentum vector. In the case of tau decays, M_T is expected to peak at very low values constrained by the small tau invariant mass while in the case of W, M_T is expected to peak in the W mass. For both $e + \tau_h$ and $\mu + \tau_h$ final states, a requirement of $M_T < 40$ Gev is applied. While transverse mass provides a good discrimination against W+jets and $t\bar{t}$, there are cases where improvement can be made in rejecting those backgrounds. In the case of $e + \mu$ where the main background is $t\bar{t}$ and the transverse mass is smeared by the presence of neutrinos in both legs, a different variable is introduced to reject this background. The variable was first introduced in MSSM Higgs searches in Tevatron [1] and it exploits the fact that the taus are boosted. Therefore the neutrinos from the tau decay are produced near the visible decay products. The bisector angle between the visible products defines the ζ axis. The visible transverse momenta and the $\not{E}_T$ are projected on the

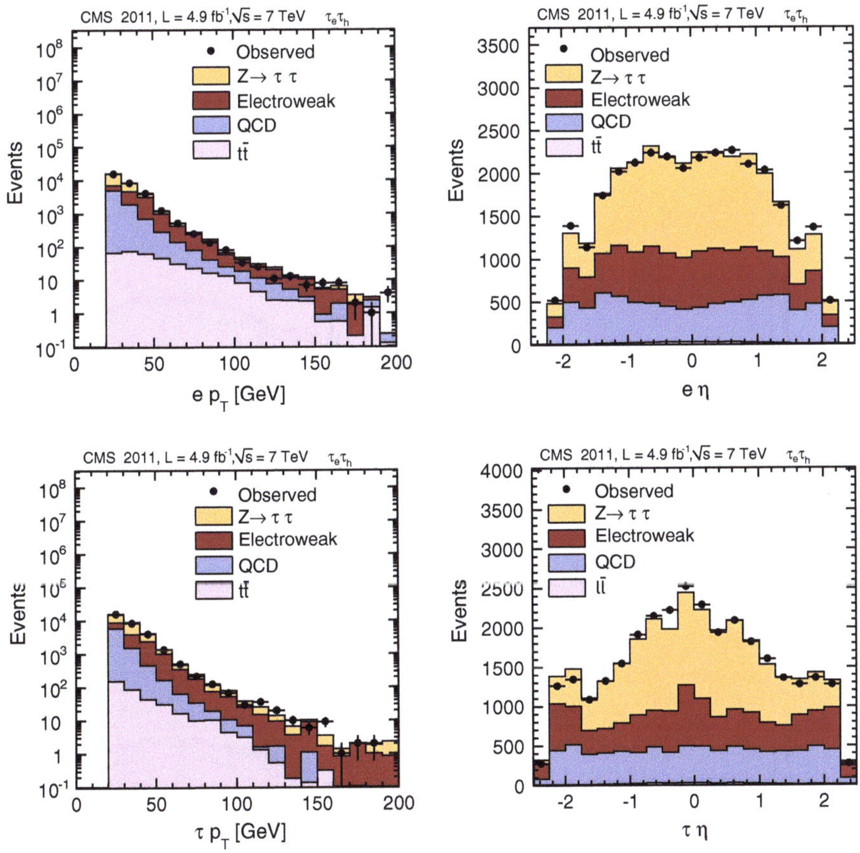

Fig. 7.2 Electron and Tau transverse momentum and pseudorapidity for the $e + \tau_h$ final state

ζ axis and two values are calculated:

$$p_\zeta = p_{\mathrm{T}}^1 \cdot \hat{\zeta} + p_{\mathrm{T}}^2 \cdot \hat{\zeta} + \not{E}_T \cdot \hat{\zeta}, \tag{7.2}$$

$$p_\zeta^{\mathrm{vis}} = p_{\mathrm{T}}^1 \cdot \hat{\zeta} + p_{\mathrm{T}}^2 \cdot \hat{\zeta} \tag{7.3}$$

Figure 7.4 shows an illustration of the p_ζ variables for a real di-tau event and a W+jets event. In the case of real di-tau event the values and directions of p_ζ and p_ζ^{vis} are similar while in the case of W+jets the opening angle between the lepton and the missing transverse energy changes the direction and values of the vectors. The collinearity requirement is imposed by using a combined variable:

$$p_\zeta^{\mathrm{cut}} = P_\zeta - \alpha \cdot P_\zeta^{\mathrm{vis}} \tag{7.4}$$

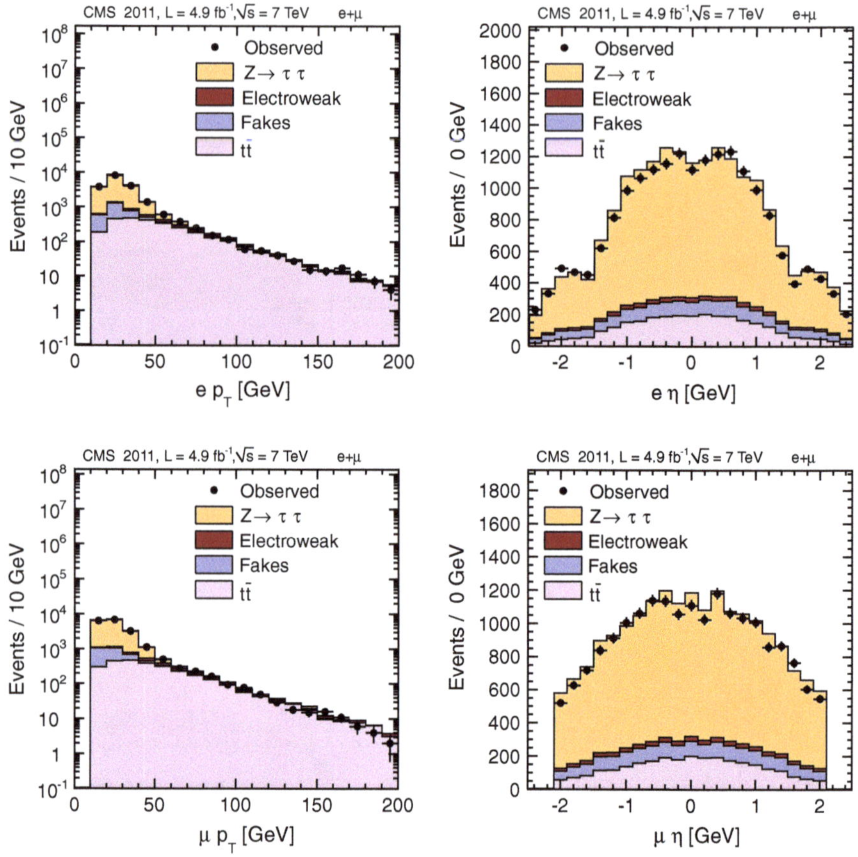

Fig. 7.3 Electron and muon transverse momentum and pseudorapidity for the $e + \mu$ final state

In the case of $e + \mu$ final state a requirement of $p_{\zeta}^{\text{cut}} > -25$ Gev is applied with $\alpha = 1.25$. The ζ variables are very useful also in the case of final states with hadronic taus when searching for high mass Higgs bosons in the MSSM. In this case the transverse mass requirement causes an efficiency loss for the signal therefore is replaced by a requirement of $p_{\zeta}^{\text{cut}} > -20$ Gev with $\alpha = 1.20$. The spectra of P_{ζ}^{cut} and M_T are shown in Fig. 7.5.

To discriminate against the $Z \rightarrow \ell^+\ell^-$ background a di-lepton veto is introduced. For the muon case, the veto requires no second global muon with $p_{\text{T}} > 15$ Gev/c, opposite charge to the first, and $I_{rel}^{PF}(\Delta\beta) < 0.2$ (Sect. 5.5). For the electron case,the veto requires no second electron with $p_{\text{T}} > 15$ Gev/c, opposite charge to the first, and $I_{rel}^{PF}(\Delta\beta) < 0.3$. For the $e + \mu$ final state, both di-lepton vetos are applied. Finally, in all final states the two legs are required to have opposite charge.

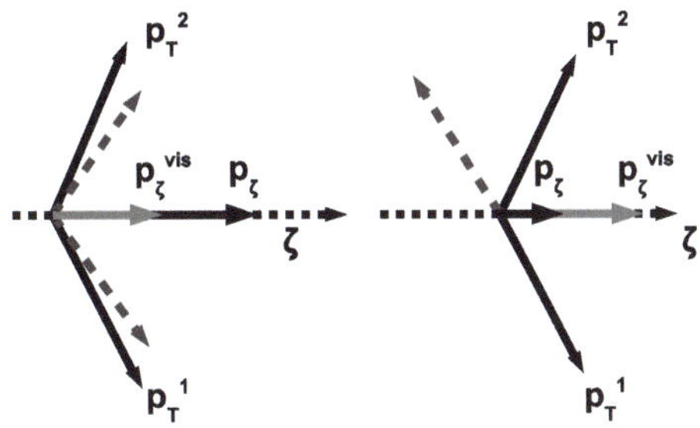

Fig. 7.4 Illustration of the ζ variables for a real di-tau event (*left*) and a W+jets event (*right*)

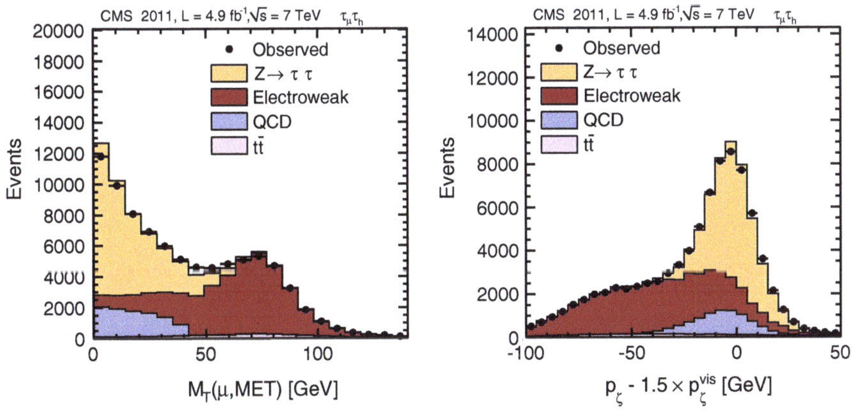

Fig. 7.5 Transverse mass and ζ variable for the $\mu + \tau_h$ final state

7.4 Corrections to the Simulation

Despite the excellent simulation of the CMS Detector, there are still small differences between the simulation and the data. Therefore, the simulation has to be corrected to make appropriate predictions. The corrections to the simulation are performed by measuring the quantities to be corrected in signal free regions in data and simulation and applying event weights to the simulated samples that correspond to the respective ratio of the quantities. The general form of the event weight for correction on objects is:

$$w_{\text{eff}} = \frac{\epsilon^{\text{Data}}(\vec{x}_i)}{\epsilon^{\text{Sim}}(\vec{x}_i)}, \tag{7.5}$$

where $\epsilon^{\text{Data}}(\vec{x}_i)$ and $\epsilon^{\text{Sim}}(\vec{x}_i)$ are the respective efficiencies in data and simulation respectively and x_i are the quantities that the efficiencies depend on, e.g. the p_T and the η of a lepton. The following paragraphs describe the corrections performed in this analysis.

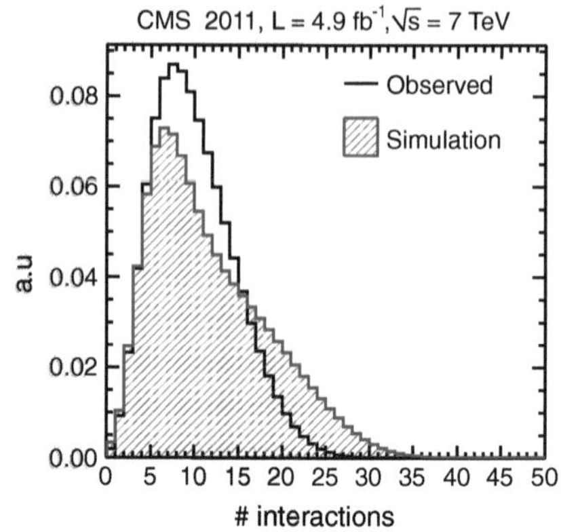

Fig. 7.6 Distribution of the number of interactions for simulation and data as extracted by the luminosity information

7.4.1 Pileup Corrections

As described in Sect. 4.2.1, the MC samples have been simulated using a uniform pileup distribution. This distribution is re-weighted to match the pileup distribution in the data. In simulation, the number of interactions for each bunch crossing is available for each event. In data, this information can be derived from the luminosity information. From Sect. 3.1.3 the number of PU interactions per crossing is:

$$\mu = \frac{\mathcal{L}\sigma_T}{R_B f_B}. \tag{7.6}$$

Therefore, knowing the revolution frequency R_B, the fraction of filled bunches (f_B) and the total pp cross section, one can get the average number of interactions for each lumisection. A lumisection is defined as the time needed for 2^{18} orbits and it is the minimum time segment where the instantaneous luminosity is calculated. The PU distribution in data is derived using the information about the number of interactions from each lumisection and using the list of lumisections used in the analysis. Then the simulated distribution is re-weighted to match the observed distribution by assigning event weights. Figure 7.6 shows the obtained distributions from data and simulation. For the data distribution, the total pp cross section used is the one obtained by the TOTEM experiment [2] and corresponds to $\sigma_T = 73$ mb.

7.4.2 Electron and Muon Efficiency Corrections

To estimate the efficiency of identification and trigger, $Z \rightarrow \ell\ell$ events are used. The method applied to measure the efficiency is known as the "tag-and-probe" method. Clean Z events are selected by applying tight identification requirements to the leading lepton. In the case of muons, it is required to have $p_{\mathrm{T}} > 25$ Gev, be within

$\eta < 2.1$ and satisfy the tight identification criteria described in Sect. 5.3. In addition, the muon is required to be isolated using $I_{\ell}^{PF/\Delta\beta} < 0.2$. In the case of electrons, the electron is required to have $p_{\mathrm{T}} > 25$ Gev, be within $\eta < 2.1$ and satisfy the identification criteria described in Sect. 5.2. Isolation is applied on the lead electron, requiring $I_{\ell}^{PF/\Delta\beta} < 0.2$. The other leg of the Z candidate is then used as a probe to evaluate the identification and trigger requirements. The muon efficiency is factorized in the following steps:

$$\epsilon_{\mu} = \epsilon_{trk} \times \epsilon_{ID} \times \epsilon_{HLT}, \tag{7.7}$$

where ϵ_{trk} is the efficiency to reconstruct a track in the inner tracker and ϵ_{ID} is the efficiency to reconstruct a full muon passing all the requirements described in 5.3. To evaluate ϵ_{trk}, Z events are selected by requiring a high quality muon satisying the criteria discussed in Sect. 5.3 and a standalone muon track that has opposite charge to the muon. Whereas, to evaluate the identification efficiency, a high quality muon and an opposite sign inner track are required. To increase the purity of Z events, in both cases a mass window of $60 < M_{\mathrm{tp}} < 120$ Gev is required.

The electron efficiency is factorized as following:

$$\epsilon_{e} = \epsilon_{SC} \times \epsilon_{ID} \times \epsilon_{HLT}, \tag{7.8}$$

where ϵ_{SC} is the probability to reconstruct a supercluster and ϵ_{ID} is the probability to reconstruct a GSF track and requiring that the electron object criteria described in 5.2 have passed. To measure the GSF track and identification efficiency, an electron plus supercluster candidate is formed while to measure the supercluster efficiency an electron plus track candidate is formed. Similarly to the muon case, a mass window around the Z resonance is applied by requiring $60 < M_{\mathrm{tp}} < 120$ Gev. For both electrons and muons, the trigger efficiency ϵ_{HLT} is defined as the probability that the probe muon passes the respective HLT path when it has passed all identification and isolation requirements. The efficiency is measured by defining two regions: a region of events passing the respective requirement (Pass) and a region of events failing the requirement (Fail). The efficiency is defined as

$$\epsilon = \frac{N_{\mathrm{Pass}}}{N_{\mathrm{Pass}} + N_{\mathrm{Fail}}}, \tag{7.9}$$

where N_{Pass} is the number of Z events in the Pass region and N_{Fail} is the number of Z events failing the requirement.

Despite the high purity of the Z sample there is non-negligible contamination of backgrounds under the Z peak, especially in the fail regions. Therefore to extract the number of Z events a fit is performed taking into account the background contributions. The main background contributions include $Z \rightarrow \tau\tau$, W+jets, QCD and $t\bar{t}$. The Likelihood is formed as following for the Pass and fail regions respectively:

$$\mathcal{L}^P = \left[N_Z \epsilon f_Z^P(m_{\ell\ell}) + \sum_i N_i^P f_i^P(m_{\ell\ell}) \right] \times \prod_i Ln(N_i^P, \bar{N_i^P}, \sigma_i) \tag{7.10}$$

$$\mathcal{L}^F = \left[N_Z(1-\epsilon) f_Z^F(m_{\ell\ell}) + \sum_i N_i^F f_i^F(m_{\ell\ell}) \right] \times \prod_i Ln(N_i^F, \bar{N_i^F}, \sigma_i), \tag{7.11}$$

where N_Z is the number of Z events, ϵ is the efficiency and f_i are shape PDFs of the mass distributions for Z and backgrounds. To improve the fit stability a constraint to the backgrounds is applied in a form of a product of Lognormal distributions which is expressed by multiplying the Pass and Fail likelihoods with a constrain term:

$$\mathcal{L}_b kg = \prod_i^N \frac{1}{b_i \delta b_i \sqrt{2\pi}} \exp - \frac{(\ln b_i - 1)^2}{2\delta\beta_i{}^2}. \tag{7.12}$$

where b_i is the background yield of the ith background and δb_i is equal to $1+\sigma_i$ where σ_i is the uncertainty which is taken to be 30 %. The fit is performed simultaneously in the Pass and Fail regions and the efficiency is extracted.

The $Z \rightarrow \ell\ell$ signal is modeled by the line-shape derived by simulated events in the pass and fail regions respectively. To account for scale and resolution differences between data and simulation the line-shape from simulation is convolved with a Gaussian distribution. The extracted mean and width of this distribution provides additional information about the scale and resolution of the muons. The reconstruction and identification efficiencies are measured separately in the barrel and endcap and in specific regions defined by the analysis. The trigger efficiency is measured in bins of reconstructed lepton transverse momenta and the turn on is modeled by a function that is defined as the integral of the Crystal Ball [3] function given by:

$$\epsilon(p_{\mathrm{T}}) = N \cdot \begin{cases} \int dx \cdot exp\left(-\frac{(x-m_0)^2}{2\sigma^2}\right) & \text{for} \frac{x-m_0}{\sigma} > -\alpha \\ \int dx \left(\frac{n}{|\alpha|}\right)^n \cdot exp\left(-\frac{|\alpha|^2}{2}\right) \cdot \left(\frac{n}{\alpha} - |\alpha| - \frac{x-m_0}{\sigma}\right)^{-n} & \text{for} \frac{x-m_0}{\sigma} < -\alpha \end{cases} \tag{7.13}$$

The trigger efficiency results are presented in Fig. 7.7. In the case of muons, the Crystal Ball integral is replaced with a linear function at high p_{T} in the barrel region to model the dropping efficiency behavior. The identification and isolation correction factors are presented in Table 7.2 for all final states.

7.4.3 Tau Efficiency Corrections

The efficiency of the tau identification is measured in the combined fit of the three final states to extract the $Z \rightarrow \tau\tau$ cross section or a Higgs limit. However, an independent measurement that is described elsewhere [4] provides a correction factor of $\rho_{\mathcal{T}} = 1.00 \pm 0.06$ that is used as an initial constraint in the fits described in the

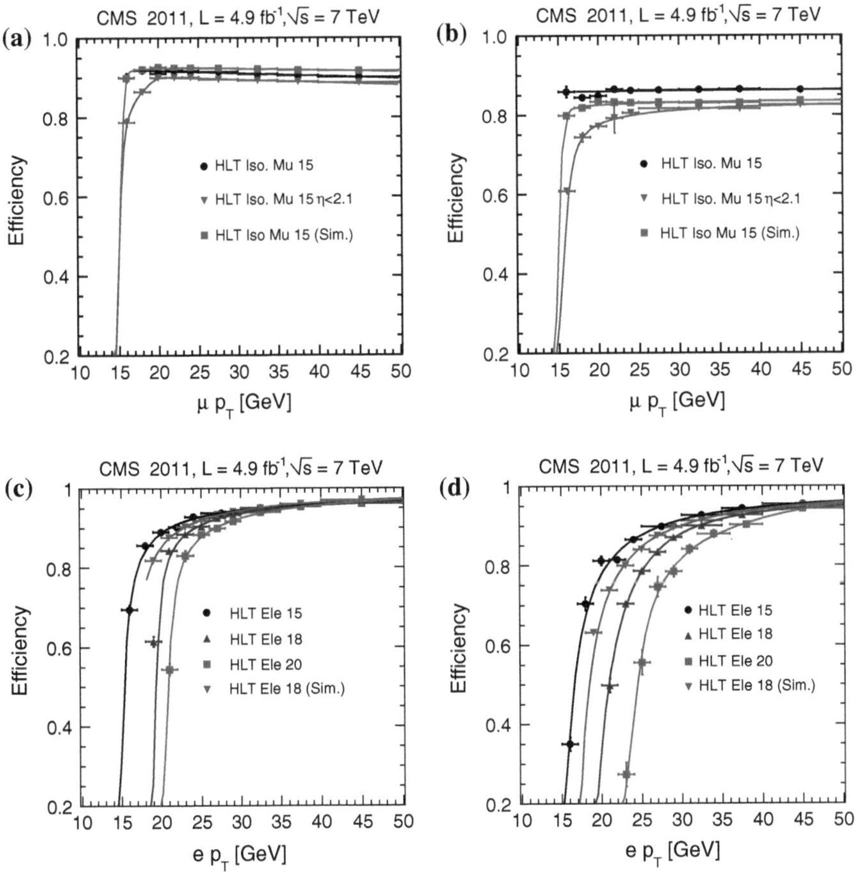

Fig. 7.7 Muon (*up*) and Electron (*down*) trigger efficiency as measured in data and simulation for barrel (*left*) and endcap (*right*) for the triggers used in the $\mu + \tau_h$ and $e + \tau_h$ final states, **a** muon HLT, barrel **b** muon HLT, **c** endcap electron HLT, barrel **d** electron HLT, endcap

following chapters. This measurement was obtained using a "tag-and-probe" like procedure but using $Z \rightarrow \tau\tau$ events. A more precise tau identification correction factor is extracted in the $Z \rightarrow \tau\tau$ cross section but since the sample of events that is used to derive the factor is the same as the one that is used to search for the Higgs, this factor is not used for the Higgs searches described in the last chapters. The efficiency of the tau trigger is measured by applying the full $\mu + \tau_h$ and $e + \tau_h$ selection in data samples that have passed the single lepton triggers and checking if the selected events also pass the cross triggers. In the case of $e + \tau_h$ final state, a set of prescaled single electron triggers was designed explicitly for this purpose. The main characteristic of the trigger efficiency measurement using $Z \rightarrow \tau\tau$ events is the background contamination from QCD and W+jets events. Those contribute fake taus that have slightly different trigger efficiency from real taus. To account for this

Table 7.2 Correction factors for the selection and identification of muons and electrons for the requirements applied in each specific final state

	Muon		Electron	
	Barrel	Endcap	Barrel	Endcap
	$\mu + \tau_h$ and $e + \tau_h$ final states			
$17 < p_T \leq 20$	0.952 ± 0.006	0.930 ± 0.008	–	–
$20 < p_T$	0.967 ± 0.006	0.943 ± 0.008	0.972 ± 0.002	0.969 ± 0.003
	$e + \mu$ final state			
$10 < p_T \leq 15$	0.98 ± 0.01	0.96 ± 0.01	0.93 ± 0.01	0.91 ± 0.01
$15 < p_T \leq 20$	0.982 ± 0.003	0.971 ± 0.007	0.948 ± 0.006	0.942 ± 0.007
$20 < p_T$	0.987 ± 0.001	0.978 ± 0.002	0.972 ± 0.003	0.967 ± 0.004

effect, an additional systematic uncertainty is assigned to account for the impurity of the sample and it is based on the trigger efficiency difference in the simulation, if the background is varied by one standard deviation of its uncertainty corresponding to approximately 10 % of the yield. The uncertainty of the background is estimated by data driven methods discussed in Sect. 7.5. Figure 7.8 shows the efficiency of the tau triggers as a function of the reconstructed tau transverse momentum in the barrel and the endcap region of the detector. The systematic uncertainty due to the background contamination is estimated to be less than 3 %.

7.4.4 Missing E_T Calibration

The missing transverse energy is a useful quantity for rejecting W+jets and $t\bar{t}$ events through the M_T or ζ requirements. To evaluate properly the shape of the background, the missing transverse energy response and resolution in the simulation has to be calibrated to match the data. The $\not{E}_T$ in this analysis is calibrated using the procedure here [5] using $Z \rightarrow \mu\mu$ events. The recoil in the Z event is defined in the transverse plane as

$$\vec{R} = -\not{E}_T - p_T^Z, \quad (7.14)$$

where p_T^Z is the transverse momentum of the Z system. The recoil is decomposed into two coefficients, one parallel (U_1) and one perpendicular (U_2)to the Z boson direction. The projected variables are fitted in bins of Z boson p_T. Response functions are created from the means of the fit results for respective bins and resolution functions from the respective widths. The resolution and response are parametrized as a function of the Z boson p_T. Using the values derived from the Z study , several samples are be calibrated. For example, W simulation is calibrated by applying the corrections as a function of the generated W boson p_T. Similar procedure is used for the Higgs samples. Since the production mechanism is different one needs to take

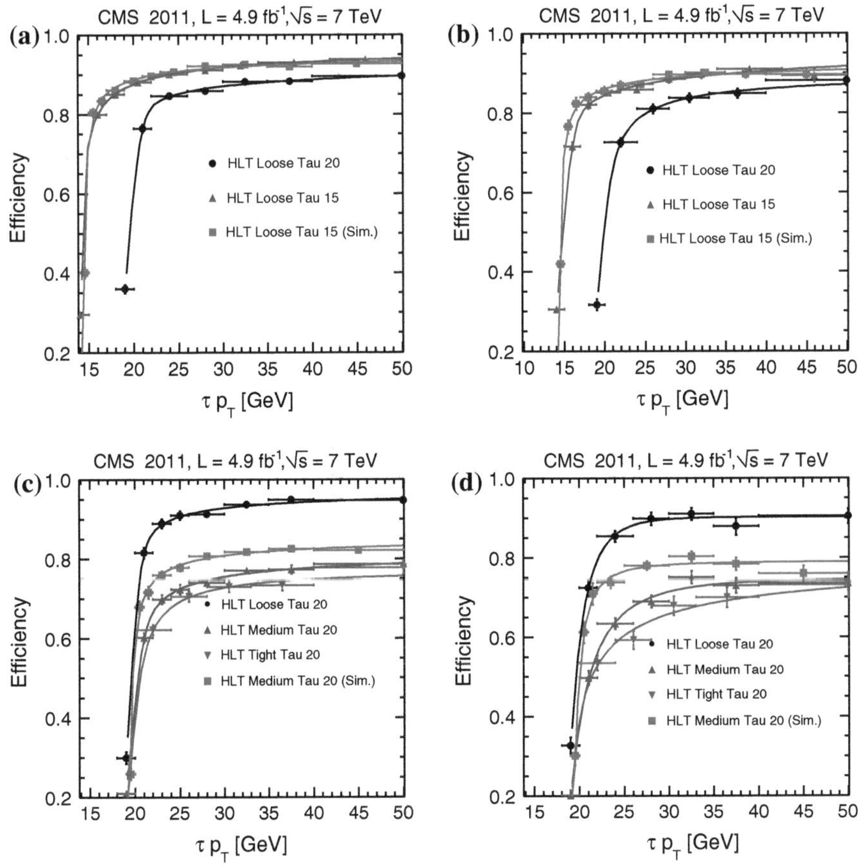

Fig. 7.8 Tau Trigger efficiency as measured in data and simulation, **a** $\mu + \tau_h$, barrel **b** $\mu + \tau_h$, endcap **c** $e + \tau_h$, barrel **d** $e + \tau_h$, endcap

into account the difference in the recoil between Higgs and Z as a systematic uncertainty. The calibration is performed simultaneously on the response and resolution. The response is corrected by applying a scale factor to the $i$the component of the recoil:

$$K_i = \frac{f_i^{\mathrm{DATA}}(p_{\mathrm{T}})}{f_i^{\mathrm{MC}}(p_{\mathrm{T}})}, \tag{7.15}$$

where f_i are the response function derived in data and simulation respectively.

For the resolution a Gaussian smearing is performed with a width equivalent to

$$\sigma_i = \sqrt{f_{i,\mathrm{DATA}}(p_{\mathrm{T}})^2 - f_{i,\mathrm{MC}}(p_{\mathrm{T}})^2} \tag{7.16}$$

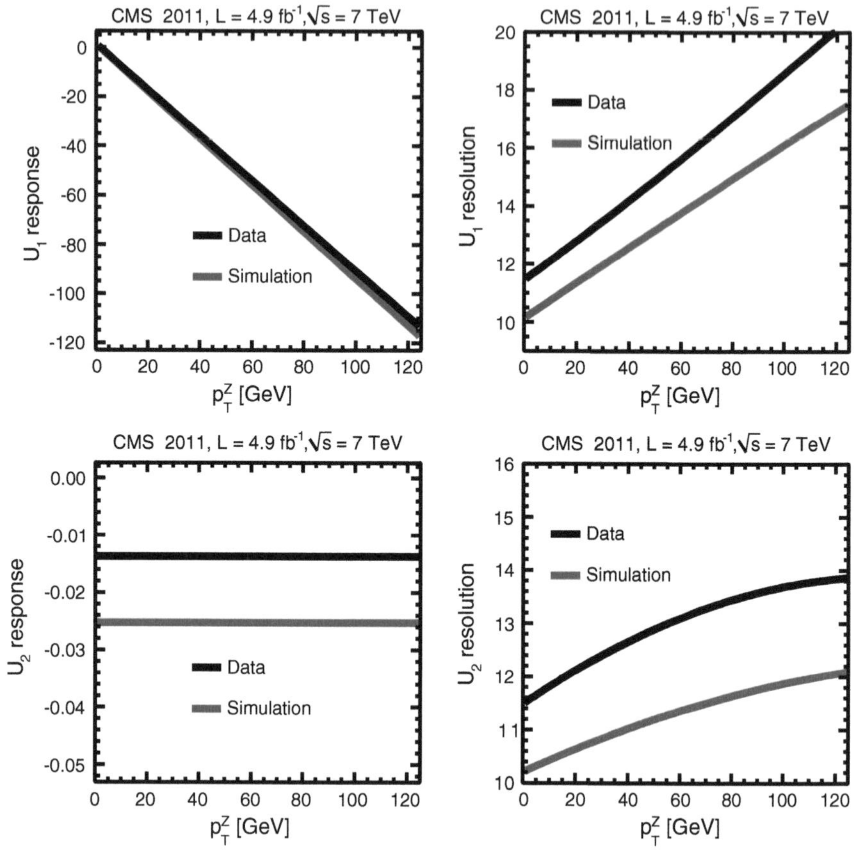

Fig. 7.9 Parametrization of the response and resolution of the recoil components for data and simulation

After the recoil has been calibrated, the missing transverse energy is recalculated with the new recoil vector for each simulated event. Figure 7.9 shows the parametrization of the recoil response and resolution as a function of the transverse momentum of the Z boson. PU re-weighting has been applied to the simulation. As expected, the response of the parallel component (U_1) is linear with the transverse momentum of the Z while the perpendicular component is insensitive to the Z transverse momentum. Both components in the simulation need additional smearing to match the data.

7.4.5 Corrections to the Topological Requirements

The efficiency of the topological requirements, namely the transverse mass, the opposite charge requirement and the di-lepton veto, are calculated for the $Z \rightarrow \tau\tau$ both in the case where it is treated as a signal ($Z \rightarrow \tau\tau$ cross section measurement) and in the case where it is treated as background (Higgs searches).

Table 7.3 Efficiency and correction factors for the topological requirements for all three final states estimated using the embedded sample

Final state	Efficiency (DATA)	Efficiency (MC)	Correction factor
$\mu + \tau_h$	0.885 ± 0.001	0.884 ± 0.002	1.001 ± 0.002
$e + \tau_h$	0.857 ± 0.002	0.857 ± 0.003	1.000 ± 0.004
$e + \mu$	0.945 ± 0.001	0.945 ± 0.002	1.000 ± 0.002

The correction factors to the topological requirements are estimated using the embedding technique in $Z \rightarrow \mu\mu$ events [4]. A clean $Z \rightarrow \mu\mu$ sample is selected and the muons are replaced with tau leptons with the muon kinematics that are allowed to decay using TAUOLA [6]. The embedded event is reconstructed and is mixed with the original $Z \rightarrow \mu\mu$ event by combining the PF candidate collections after removing the muons from Z. Then taus are reconstructed and the full analysis is repeated on the new sample. The hybrid events have the underlying event, PU and jet multiplicity from the data while the tau decay is modeled very well by TAUOLA. The efficiency of the topological requirements is measured in the embedded sample and compared to the MC simulation. Results are presented in Table 7.3.

7.5 Background Estimation in $e + \tau_h$ and $\mu + \tau_h$ Final States

The most important irreducible background contributions in the presence of hadronic taus in the final state are QCD multijet and W/Z+jets. Smaller backgrounds include $t\bar{t}$ and diboson production. Data driven methods were implemented to predict the residual contributions of those backgrounds after all selection requirements. The main characteristics of the backgrounds and the estimation methods are described in the next paragraphs.

7.5.1 QCD Multijet

In the muon case, heavy flavour decays and decays in flight produce energetic muons that pass the muon identification and a jet in the event can fake the tau signal. In the electron case, the effect of heavy flavor decays of quarks is not dominant for the electron background and the major contribution comes from fake electrons produced by jets. The main characteristic of the QCD background is that it can produce fake tau pairs with both opposite sign(OS) and same sign(SS) topologies. In the case of pure fake events, the OS/SS ratio should be compatible with unity, however, this is not the case for real pairs from heavy flavor (especially b) quark decays. The key behind

QCD measurement is estimation of the events in the OS (signal) region by using events in the SS region. To perform this extrapolation the OS/SS ratio is measured with the data.

The OS/SS ratio is measured by inverting the requrements on the lepton isolation while keeping the hadronic tau isolated. The isolation requirement in the electron trigger requires using prescaled triggers to calculate the ratio in the electron case which results in limited statistics. Figure 7.10 shows the OS and SS events in data for the inverted isolation for both final states. A slight excess is observed for OS events due to the presence of heavy flavor quark decays to real leptons. Figure 7.11 shows the OS/SS ratio as a function of the inversion requirement for the lepton isolation. The value is taken from the flat region at higher inversion values and a systematic uncertainty is applied based on the maximum deviation from the mean. The values obtained for both final states are:

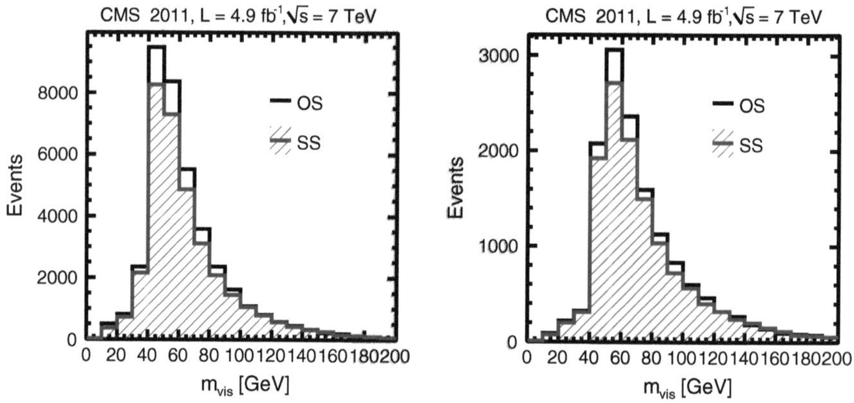

Fig. 7.10 Number of OS and SS events in the region defined by inverted lepton isolation for $\mu + \tau_h$ (*left*) and $e + \tau_h$ (*right*) final states respectively

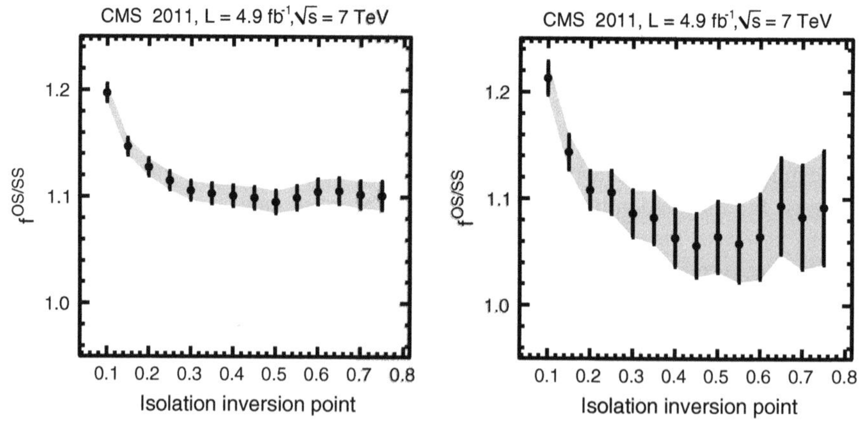

Fig. 7.11 Estimation of the OS/SS fraction in anti-isolated events as a function of the isolation requirement for $\mu + \tau_h$ (*left*) and $e + \tau_h$ (*right*) final states

$$f^{OS/SS}_{\mu+\tau_h} = 1.10 \pm 0.01(\text{stat}) \pm 0.10(\text{syst}) \tag{7.17}$$

$$f^{OS/SS}_{e+\tau_h} = 1.08 \pm 0.3\ (\text{stat}) \pm 0.15\ (\text{syst}) \tag{7.18}$$

7.5.2 W+Jets

In the case of W events, there is a real high quality isolated lepton from the W decay and a jet faking the hadronic tau. A characteristic of the W sample is an excess in OS contribution with respect to the SS contribution, therefore this background should be estimated separately and subtracted from the SS QCD events before applying OS/SS scaling factors which were discussed in the Sect. 7.5.1. Events are counted in the region $M_T > 70$ Gev, which is W enriched, for both OS and SS cases and is used for extrapolation into the signal region ($M_T < 40$ Gev). The extrapolation factor is taken from simulation where the missing transverse energy is calibrated using the recoil method. W decays can also contribute to the selected sample via $W \rightarrow \tau\nu$, where tau decays to electrons or muons. Due to the tau decay branching fraction and the smaller acceptance due to the tau decay, those events are a very small fraction of the total W+jets events. Since the production mechanism is the same the $W \rightarrow \tau\nu$ events are extrapolated simultaneously with the $W \rightarrow \ell\nu$ events. The extrapolation factors for the two final states are calculated to be:

$$f^{W+\text{jets}}_{\mu+\tau_h} = 0.379 \pm 0.028 \tag{7.19}$$

$$f^{W+\text{jets}}_{e+\tau_h} = 0.348 \pm 0.036 \tag{7.20}$$

The error is estimated by quantifying the change of this factor due to variation of the lepton energy scales as well as an additional 1 % error for the $\not{E}_T$ calibration.

7.5.3 $Z \rightarrow \ell\ell$+Jets

The $Z \rightarrow \ell\ell$ background enters in the final selection in two ways:

- A lepton fakes the tau. In this case, the shape of the visible mass is compatible with the Z lineshape and contributes mainly to OS events.
- A jet fakes the tau. In this case, one lepton is lost out of the acceptance or was not reconstructed and there is a jet in the event faking a tau. This contribution is expected to be produced equally in the OS and SS data since any of the two leptons from Z can be lost.

This background is very small and it is taken from simulation normalized to the theoretical cross section [7]. The simulation is corrected for lepton efficiency and tau fake rate. The uncertainties of the fake rates were calculated in Chap. 6.

7.5.4 Top Pairs, di-bosons and γ+Jets

Top quark pair production is an irreducible background due to the presence of energetic leptons and taus from $W \rightarrow \ell\nu$ in top decays. In addition, semileptonic top decays where there are four or more jets in the event can enter the final selection when a jet is faking a hadronc tau. The number of expected $t\bar{t}$ events is taken from the simulation normalized using the CMS measured cross section value [8] of:

$$\sigma_{t\bar{t}} = 165.8 \pm 2.2\ \text{(stat.)} \pm 10.6\ \text{(syst.)} \pm 7.4\text{(lumi.)pb} \tag{7.21}$$

Since the sample of the measurement and the sample used in this analysis is identical, the luminosity uncertainty is not relevant. Therefore, the total cross section uncertainty is 6.6 %. The photon + jets background is relevant only for the $e + \tau_h$ final state. It is a small contamination and since the photon converts to two electrons, it is expected to be OS/SS blind. In the current background estimation, γ + jets is taken into account the same way as the QCD background. Dibosons contribute mainly via WW production but there are small contributions from WZ and ZZ as well. The uncertainty assigned to the diboson samples is 30 %.

7.5.5 Background Estimation Method

For the calculation of the background sources, the transverse and OS requirements are dropped. Initially the $W \rightarrow \ell\nu$ contributions are estimated by counting the events in the sideband defined by $M_T > 70$ Gev and applying the extrapolation factor to the signal region. The calculation is performed individually for OS and SS events. For both OS and SS events, the high M_T contribution of $t\bar{t}$ and di-boson events are subtracted before extrapolation. The number of W events in sideband is calculated as:

$$W^{OS}_{sideband} = N^{OS}_{W} - t\bar{t}^{OS}_{W} - VV^{OS}_{W} \tag{7.22}$$

$$W^{SS}_{sideband} = N^{SS}_{W} - t\bar{t}^{SS}_{W} - VV^{SS}_{W} \tag{7.23}$$

The W contribution to the signal is calculated by applying the extrapolation factor using the W shape :

$$W^{OS}_{signal} = W^{OS}_{sideband} \cdot f^{W+jets} \tag{7.24}$$

Table 7.4 Background estimation results for the $\mu + \tau_h$ and $e + \tau_h$ final states

	OS		SS	
Source	$M_T < 40$ Gev	$M_T > 70$ Gev	$M_T < 40$ Gev	$M_T > 70$ Gev
$\mu + \tau_h$				
Di-Boson	212 ± 66	325 ± 200	–	–
$t\bar{t}$	813 ± 80	1813 ± 128	–	–
Z^{l+jet}	466 ± 51	–	441 ± 49	–
Z^{ll}	354 ± 108	–	28 ± 9	–
$W + jets$	5883 ± 435	15511 ± 215	2149 ± 159	5948 ± 77
QCD	8234 ± 773	–	7418 ± 195	–
$Z \to \tau\tau$	30402 ± 2107	–	–	–
Total Expected	46364 ± 2292	–	–	–
Observed	46244	17650	10188	5948
$e + \tau_h$				
Di-Boson	88 ± 27	156 ± 48	–	–
$t\bar{t}$	350 ± 37	876 ± 68	–	–
Z^{l+jet}	1672 ± 179	–	488 ± 54	–
Z^{ll}	3803 ± 404	–	255 ± 29	-
$W + jets$	3172 ± 333	9132 ± 130	1122 ± 118	3365 ± 58
QCD	8557 ± 1202	–	7923 ± 165	–
$Z \to \tau\tau$	13827 ± 1081	–	–	–
Total Expected	31467 ± 1709	–	–	–
Observed	30679	10164	9853	3365

All backgrounds are estimated with the methods described in the text. For the $Z \to \tau\tau$ contribution, the events are normalized using the theoretical cross section [7]

$$W^{SS}_{signal} = W^{SS}_{sideband} \cdot f^{W+jets} \tag{7.25}$$

QCD is first estimated in the SS region after subtracting all other other SS contributions and is given by:

$$QCD^{SS}_{signal} = N^{SS}_{signal} - W^{SS}_{signal} - Z^{SS}_{\ell+jet} - Z^{SS}_{\ell\ell} - t\bar{t}^{SS} - VV^{SS} \tag{7.26}$$

and the OS QCD is calculated using the OS/SS factor:

$$\begin{aligned} QCD^{OS}_{signal} &= f^{OS/LS} \cdot QCD^{SS}_{signal} \\ &= f^{OS/LS} \cdot \left(N^{SS}_{signal} - W^{SS}_{signal} - Z^{SS}_{\ell+jet} - Z^{SS}_{\ell\ell} - t\bar{t}^{SS} - VV^{SS}\right) \end{aligned} \tag{7.27}$$

The number of extracted events are summarized in Table 7.4 for both final states.

7.6 Background Estimation for the $e + \mu$ Final State

The $e+\mu$ final state is characterized by dominant $Z \rightarrow \tau\tau$, $t\bar{t}$ and backgrounds from fake leptons (QCD, W/Z+jets). Other smaller sources include di-bosons where irreducible WW production dominates. The $t\bar{t}$ and diboson backgrounds are estimated by normalizing the simulation to the measured values of the respective cross sections and applying the proper correction factors.

Fake backgrounds are estimated using data by exploiting the following regions depending on the charge of the di-tau candidate and the lepton isolation.

- A: OS, both leptons isolated.
- B: OS, both leptons anti-isolated.
- C: SS, both leptons isolated.
- D: SS, both leptons anti-isolated.

Regions A and C are dominated by fake backgrounds while region C has small contamination from $t\bar{t}$ and dibosons. The uncertainty of those backgrounds is propagated to the uncertainty of the events in the region. In the signal free SS regions, an extrapolation factor is defined as

$$f^{\text{fake}} = \frac{N_C}{N_D}, \tag{7.28}$$

where N_C, N_D is the number of events in C and D region respectively, after subtracting small backgrounds from the C region. The extrapolation factor is then applied on the number of events in the B region to give the expected fake background in region A which is given by:

$$N_A^{\text{fake}} = N_B \times f^{\text{fake}} = \frac{N_B N_C}{N_D} \tag{7.29}$$

Since it is required that both leptons are not isolated in the inverted regions the estimation is dominated by QCD. Other minor fake contamination is due to a vector boson plus a jet faking a tau. Those contributions are estimated using the simulation

Table 7.5 Background estimation results for the $e + \mu$ final state

Source	OS	SS
$e + \mu$		
Di-Boson	428 ± 129	26 ± 8
$t\bar{t}$	2489 ± 201	23 ± 3
Fakes	1560 ± 178	751.00 ± 50
$Z \rightarrow \tau\tau$	14054 ± 739	–
Total Expected	18530 ± 797	800 ± 51
Observed	18316	848

All backgrounds are estimated with the methods described in the text. For the $Z \rightarrow \tau\tau$ contribution, the events are normalized using the theoretical cross section [7]

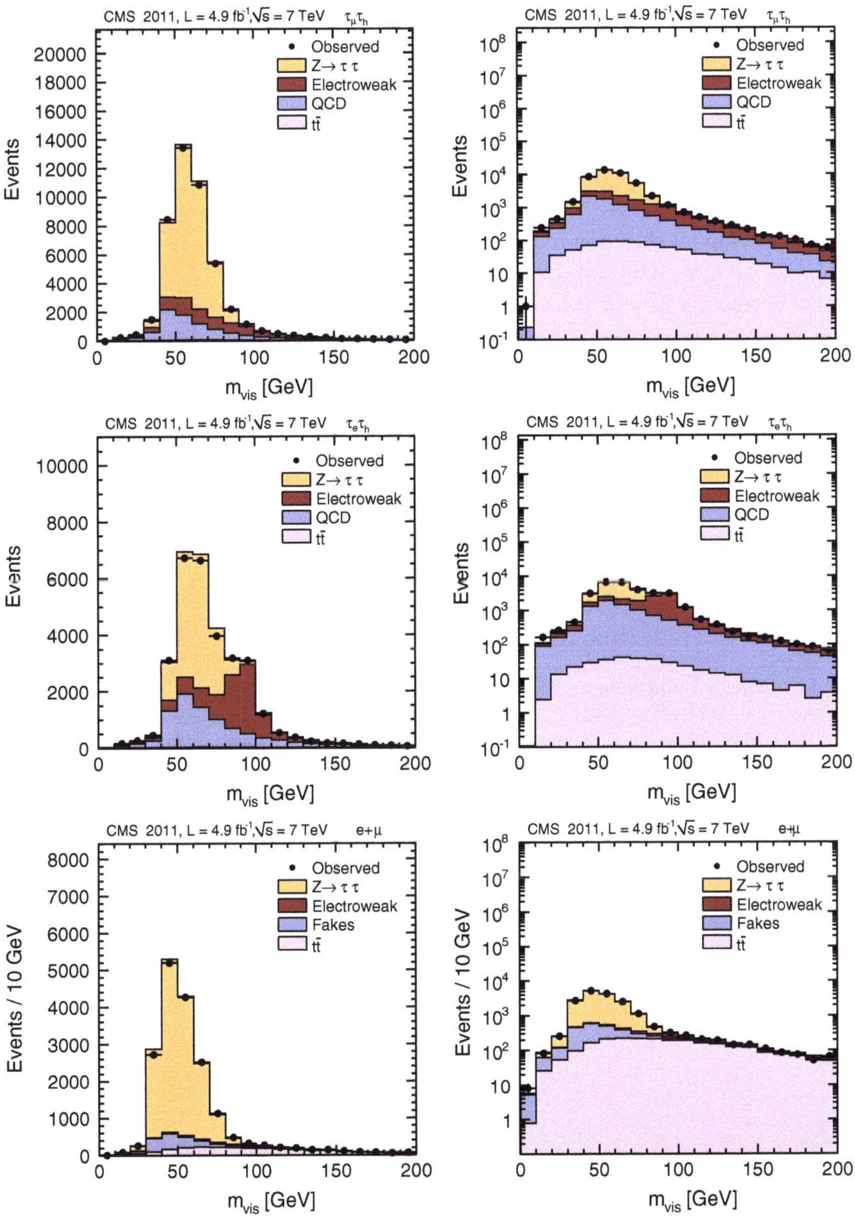

Fig. 7.12 Visible mass distribution of all three final states in linear (*left*) and logarithmic scale (*right*) after complete background estimation and correction factors applied to the simulation

assigning a conservative uncertainty of 40 % Table 7.5 shows the results of the data driven estimation for the $e + \mu$ final state.

7.7 Results

After applying all selection requirements a data sample of about 70 K events is selected for all three final states. After data driven background estimation and application of correction factors, the sample is compatible with SM expectation compatible with $Z \to \tau\tau$ production. Figure 7.12 shows the visible mass distributions in data compared to simulation after all corrections have been applied.

References

1. C.C. Almenar, Search for the neutral MSSM Higgs bosons in the tau tau decay channels at CDF Run II. Ph.D. thesis (2008)
2. G. Antchev, P. Aspell, I. Atanassov, V. Avati, J. Baechler et al., First measurement of the total proton-proton cross section at the LHC energy of sqrt(s) =7 TeV. Europhys. Lett. **96**, 21002 (2011)
3. J.E. Gaiser, Charmonium spectroscopy from radiative decays of the J/Psi and Psi-prime. Ph.D. thesis, SLAC-R-255 (1982), p. 178
4. The CMS Collaboration, Search for Neutral Higgs bosons Decaying to Tau Pairs in pp Collisions at $\sqrt{s} = 7$ TeV. *CMS Physics Analysis Summary*, CMS-PAS-HIG-11-029 (2011)
5. P.Harris et al., Modeling of $W \to \ell\nu$ missing transverse energy with boson recoil. CMS-AN-2010-332 (2010)
6. Z. Was, TAUOLA the library for tau lepton decay. Nucl. Phys. Proc. Suppl. **98**, 96–102 (2001)
7. R. Gavin, Y. Li, F. Petriello, S. Quackenbush, FEWZ 2.0: a code for hadronic Z production at next-to-next-to-leading order (2010). http://arxiv.org/abs/1011.3540
8. CMS Collaboration, Combination of top pair production cross section measurements. *CMS Physics Analysis Summary*, CMS-PAS-TOP-11-024 (2011)

Chapter 8
Measurement of $Z \to \tau\tau$ Production

This chapter presents a measurement of the Z production cross section when Z decays to tau pairs using the 2011 data sample. The analysis is performed in all three final states using the object reconstruction and selection requirements presented in previous chapters. The cross section is extracted using the formula:

$$\sigma = \frac{N - N_b}{\mathcal{BR} \cdot A \cdot \epsilon \cdot L}, \tag{8.1}$$

where $\mathcal{BR}$ is the branching ratio of the di-tau to the final state of interest, N is the number of events in data, N_b is the number of estimated background events, A is the acceptance ϵ is the efficiency and L is the integrated luminosity. The product of acceptance and efficiency can be expressed in terms of the correction factors as estimated in Sect. 7.4 as following:

$$\bar{A} = A \times \epsilon_{MC} \times \prod_i \rho_i, \tag{8.2}$$

where ϵ_{MC} is the efficiency in the simulation and $\prod \rho_i$ is the product of all correction factors.

8.1 Acceptance Model

The acceptance is defined as the fraction of the $Z \to \tau\tau$ events that are produced within the detector and satisfy some fiducial requirements on the transverse momenta of the visible decay products. The acceptance is calculated using POWHEG [1] simulation interfaced with TAUOLA [2]. Due to the peculiarity of the tau decay to invisible products, it is not possible to define the acceptance in the Z mass window. Therefore, the acceptance is defined for $M_{\tau\tau} > 20\,\text{GeV}$ using the visible decay products. However, to compare the cross section with the one derived for other

M. Bachtis, *Heavy Neutral Particle Decays to Tau Pairs*,
Springer Theses, DOI: 10.1007/978-3-319-03257-3_8,

Table 8.1 Acceptance calculated with POWHEG for all three final states

Final state	Acceptance (%)
$\mu + \tau_h$	5.889 ± 0.011
$e + \tau_h$	4.524 ± 0.009
$e + \mu$	8.084 ± 0.011

final states for $M_{\ell\ell} > 50\,\text{GeV}$, the measured cross section is extrapolated to the $M_{\tau\tau} > 50\,\text{GeV}$ window using a factor derived using FEWZ NNLO cross section calculation program [3] which is estimated to be 0.584. In the case of $\mu + \tau_h$, the acceptance requires a muon or $p\text{T} > 17\,\text{GeV}$ and $\eta < 2.1$ and a hadronic tau with visible $p\text{T} > 20\,\text{GeV}$ and $\eta < 2.3$. For the case of $e + \tau_h$, the acceptance requires an electron of $p\text{T} > 20$ GeV and $\eta < 2.1$ and a hadronic tau with visible $p\text{T} > 20\,\text{GeV}$ and $\eta < 2.3$. Finally for the $e + \mu$ final state an electron-muon pair is required with asymmetric thresholds of 20 and 10 GeV respectively where the pseudorapidity of the leptons is constrained within $\eta < 2.3$ for muons and electrons. Table 8.1 shows the acceptance and the extrapolation factor $f_M(50)$ used to extrapolate the acceptance to the $M_{\tau\tau} > 50$ mass window.

8.2 Systematic Uncertainties

The major systematic uncertainties that affect this measurement are due to experimental factors, namely, the mis-modeling of the object reconstruction, identification and energy scale and resolution. Other systematic uncertainties contribute due to theoretical uncertainties on the acceptance calculation. The sources of systematic uncertainties are discussed in the following sections.

8.2.1 Experimental Systematics

The major systematic uncertainties are due to the muon, electron and tau identification and trigger efficiencies. The muon and electron identification and trigger efficiency is measured precisely using $Z \to \ell\ell$ events as described in Sect. 7.4. The statistical uncertainties from the fit are negligible and a 1 % systematic factor is applied to account for systematic effects in mis-modelling the shapes. In the case of the $e + \mu$ final state, the systematic uncertainty assigned is 2 % to account for the higher uncertainty in measuring the efficiencies in the region defined by the lower transverse momentum thresholds.

The tau identification uncertainty is estimated with a "tag-and-probe" method to be 6 %. The efficiencies for electron and muon reconstruction, identification, isolation and trigger efficiencies are obtained from data. Correction factors for the values extracted from the simulation are determined with from $Z \to \ell\ell$ events as described in

Sect. 7.4. The uncertainties on the correction factors are estimated to be below 1 %. A similar technique is used to estimate the hadronic tau reconstruction efficiency corresponding to an uncertainty of 6 %. The tau trigger uncertainty is dominated by the background contamination in the measurement region (Sect. 7.4.3) which corresponds to 3.5 %. The uncertainty of the topological requirements is estimated using the embedding technique and corresponds to less than 0.5 % due to the high statistics of the embedding sample. The tau energy scale uncertainty is estimated by confirming the agreement of the reconstructed tau mass between data and simulation and by evaluating the maximum effects on the scale due to tau decay mode misidentification. The maximum deviations were found to be well below 3 %.

8.2.2 Theoretical Uncertainties

The main source of theoretical systematic effects is due to the uncertainty from the Parton Distribution Functions (PDFs). The effect of the PDF uncertainties in the acceptance is estimated by evaluating the acceptance difference due to different PDF sets and by studying the effect of each individual PDF set. For the estimation of the uncertainty in acceptance, the recipe prescribed by the PDF4LHC working group is followed [4] yielding an uncertainty of 2 %. Other theoretical effects include initial state radiation (ISR), final state radiation (FSR) and effects from higher orders in perturbation theory. Final state radiation is negligible in the di-tau final states and the other effects are expected to modify the transverse momentum spectrum of the Z boson usually towards higher values. The Z boson spectrum is well measured using $Z \rightarrow \mu\mu$ data as presented in Fig. 8.1. Since several additional factors such as muon momentum scale and different background contamination in data and simulation could contribute to the difference of the spectra a conservative estimate is followed for the uncertainty. The POWHEG spectrum is re-weighted to match the data and the difference in the acceptance before and after re-weighting is assigned as the systematic uncertainty. After applying the weights and re-evaluating the acceptance, the maximum difference is below 0.5 % with respect to the POWHEG estimation which is assigned to the systematics.

The experimental and theoretical uncertainties are summarized in Table 8.2.

8.3 Combined Fit

The accuracy of the cross section measurement is improved by performing the measurement simultaneously in the three final states, exploiting a combined fit. In addition, the dominant systematic uncertainty of the tau identification efficiency exists in only the final states with a hadronic tau ($e + \tau_h$, $\mu + \tau_h$), and it will be constrained in the fit by comparing the yields with the $e + \mu$ final state. The combined fit is performed exploiting also shape information by fitting the visible mass shape in each

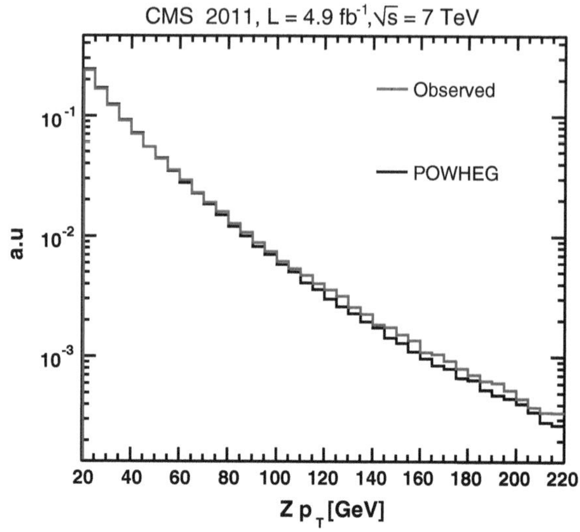

Fig. 8.1 Z transverse momentum spectrum in POWHEG simulation and $Z \to \mu\mu$ data

Table 8.2 Summary of the systematic uncertainties

Source	$\mu + \tau_h$ (%)	$e + \tau_h$ (%)	$e + \mu$ (%)
Muon identification and trigger	1	–	2
Electron identification and trigger	–	1	2
Tau identification	6	6	–
Tau trigger	3.3	3.3	–
Efficiency of topological selections	0.5	0.5	0.5
Tau energy scale uncertainty	3	3	–
Luminosity	2.2	2.2	2.2
Parton distribution functions	2	2	2
Other theoretical uncertainties	0.5	0.5	0.5

final state. The likelihood for each final state can be factorized as:

$$\mathcal{L} = \mathcal{L}_{\text{shape}} \times \mathcal{L}_{\text{bkg}} \times \mathcal{L}_{\text{syst}} \tag{8.3}$$

where $\mathcal{L}_{\text{shape}}$ is the likelihood term describing the sum of signal and background shapes normalized by their respective background estimation coefficients, $\mathcal{L}_{\text{bkg}}$ corresponds to the variation of the background estimation coefficients within their uncertainties and $\mathcal{L}_{\text{syst}}$ corresponds to the variations of the systematics factors within their uncertainties. The shape likelihood term is expressed as:

$$\mathcal{L}_{\text{shape}} = N_{Z\to\tau\tau}(n_j) f_{Z\to\tau\tau}(m_{\text{vis}}, n_j) + \sum N_i(n_j) f_i(m_{\text{vis}}, n_j), \tag{8.4}$$

where $N_{Z\to\tau\tau}(n_j)$ is the number of $Z\to\tau\tau$ events, $f_{Z\to\tau\tau}$ is the $Z\to\tau\tau$ shape PDF and $N_i(n_j)$ and f_i correspond to the number of background events and background shapes respectively. The factors n_j correspond to the vector of systematic factors(i.e. tau efficiency). The background component of the Likelihood is defined as:

$$\mathcal{L}_{\text{bkg}} = \prod_i Ln(N_i(\bar{n}_j), N_i, \sigma_i), \tag{8.5}$$

where $\bar{N}_i$ and σ_i are the mean values of the backgrounds and the respective uncertainties. Finally the systematics part of the likelihood is:

$$\mathcal{L}_{\text{syst}} = \prod_j Ln(\bar{n_j}, n_j, \sigma_j), \tag{8.6}$$

where σ_j denotes the values of the systematic uncertainties. It is possible to directly fit for the cross section by replacing

$$N_{Z\to\tau\tau}(n_j) = \sigma\bar{A}(n_j)L, \tag{8.7}$$

where σ is the cross section and L is the integrated luminosity of the sample.

The effect of the energy scale uncertainty affects both the shape and the normalization of the templates. Shape morphing is implemented to account for those effects [5]. For each scale uncertainty source, two new templates are introduced where the shape is allowed to vary up (f^+) and down (f^-) by one standard deviation of the scale uncertainty with respect to the original template (f^0). The final template is then given by an interpolation between the three depending on the scale coefficient n_s as following:

$$f(n_s) = f^0 - \alpha(n_s)\cdot f^+ + \beta(n_s)\cdot f^0 + \gamma(n_s)\cdot f^- \tag{8.8}$$

The simplest form of the α, β, γ coefficients corresponds to linear interpolation of the form $\alpha = \max(n_s, 0), \beta = -|n_s|, \gamma = \max(-n_s, 0)$. The linear model has discontinuities in the derivatives at zero, therefore for values of $n_s < 1$, quadratic interpolation is used in the following form:

$$f(n_s) = f^0 - \frac{n_s\cdot(1+n_s)}{2}\cdot f^+ + n_s^2\cdot f^0 - \frac{n_s\cdot(1-n_s)}{2}\cdot f^- \tag{8.9}$$

For each scale uncertainty source, a Lognormal factor is created in the respective yield allowing the yield of the template to vary as the scale varies which is the result of migration of events in and out of the sample.

8.4 Results

For each final state, the likelihood is minimized separately to extract the individual cross sections. The final states with a hadronic tau ($\mu + \tau_h$, $e + \tau_h$) are dominated by the tau identification efficiency uncertainty of 6 % while the $e + \mu$ final state is not affected by it. The combined fit introduces additional constraints that the channels with a hadronic tau correspond to the same cross section and their cross section be consistent with that of the $e + \mu$ final state. Moreover, the electron and muon identification efficiency is correlated between the channels that have the respective objects in the final state.

The extracted cross section for each final state extrapolated into the $M_{\tau\tau} > 50\,\text{GeV}$ window is :

$$\sigma(\mu + \tau_h) = 947 \pm 10\ (\text{stat}) \pm 66\ (\text{syst}) \pm 21\ (\text{lumi})\ \text{pb} \tag{8.10}$$

$$\sigma(e + \tau_h) = 936 \pm 17\ (\text{stat}) \pm 68\ (\text{syst}) \pm 21\ (\text{lumi})\ \text{pb} \tag{8.11}$$

$$\sigma(e + \mu) = 958 \pm 11\ (\text{stat}) \pm 34\ (\text{syst}) \pm 21\ (\text{lumi})\ \text{pb} \tag{8.12}$$

Since the fit provides a single error corresponding to both statistical and systematic contributions, the statistical part is estimated by generating and fitting toy MC experiments without the systematics part of the likelihood. Then, the systematic error is produced by subtracting in quadrature the statistical contribution. The three final states are combined using a simultaneous fit to extract the cross section and the tau identification efficiency. The result of the fit is presented in Fig. 8.2. The standard error ellipse is shown along with the 68 and 95 % confidence intervals that correspond to the simultaneous measurement of the two parameters. The 68 and 95 % CL contours are produced by varying the likelihood by $2\Delta ln\mathcal{L}$ equal to 2.30 and 5.99 respectively. The luminosity uncertainty is not shown. Considering the tau identification efficiency as a nuisance parameter and integrating it out yields a cross section for the combined measurement equal to:

$$\sigma(\tau\tau) = 955 \pm 7\ (\text{stat}) \pm 33\ (\text{syst}) \pm 20\ (\text{lumi})\text{pb} \tag{8.13}$$

The combined and the individual cross section are in good agreement with the theoretical predictions [3]. Considering the fit as a measurement of the tau identification efficiency and integrating out the cross section yields a correction factor for the tau identification equal to:

$$\rho_\tau = 0.978 \pm 0.035. \tag{8.14}$$

This correction factor proves excellent description of the hadronic taus in the simulation and a very small tau identification uncertainty of 3.5% that can be used in

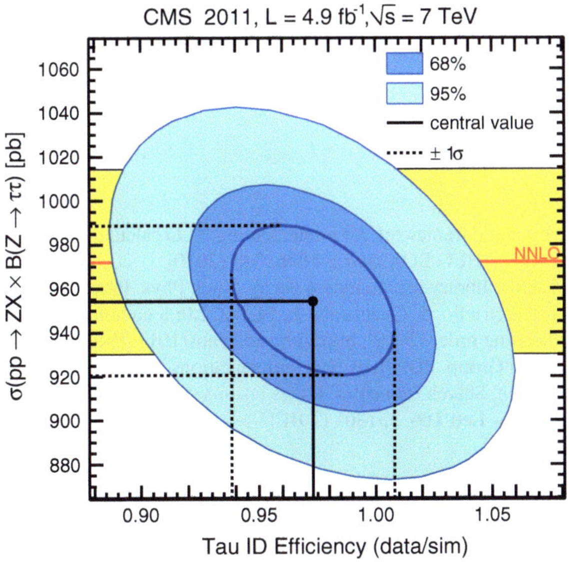

Fig. 8.2 Simultaneous measurement of the $Z \rightarrow \tau\tau$ production cross section and the tau identification efficiency correction factor. The 68 % and 95 % confidence level for the simultaneous measurement are shown as well as the standard error ellipse. The theoretical expectation at NNLO [3] is overlayed. Luminosity uncertainty of 2.2 % is not shown

Fig. 8.3 Summary of cross section measurements for the individual final states and the combination. The theoretical expectation at NNLO [3] is overlayed. Luminosity uncertainty of 2.2 % is included in the systematic error

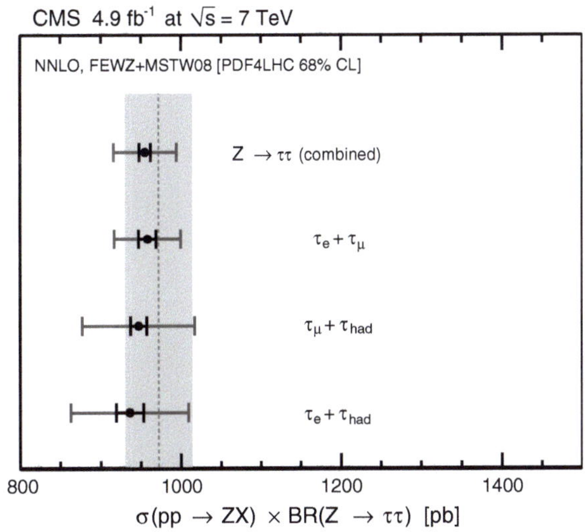

other measurements or searches with tau leptons. Figure 8.3 shows the summary of the cross section measurements for all the final states and the combination. The measurement in the $\mu + \tau_h$ and $e + \tau_h$ final states is dominated by the tau identification efficiency uncertainty while the total uncertainty of the combination is similar to the $e + \mu$ uncertainty that is not affected by tau identification.

References

1. S. Alioli et al., A general framework for implementing NLO calculations in shower Monte Carlo programs: the POWERHEG BOX. JHEP **1006**, 043 (2010)
2. Z. Was, TAUOLA the library for tau lepton decay. Nucl. Phys. Proc. Suppl. **98**, 96–102 (2001)
3. R. Gavin, Y. Li, F. Petriello, S. Quackenbush, FEWZ 2.0: a code for hadronic Z production at next-to-next-to-leading order (2010). http://arxiv.org/abs/1011.3540
4. PDF4LHC Working Group. PDF4LHC Recommendations (2010)
5. C.M.S. Collaboration, Search for neutral Higgs bosons decaying to tau pairs in PP collisions at $\sqrt{s}$=7 TeV. Phys. Rev. Lett **106**, 231801 (2010)

Chapter 9
Search for Higgs Bosons

Based on the phenomenological results of spontaneous symmetry breaking in the SM and the MSSM, both SM Higgs bosons at low mass and MSSM Higgs bosons in the whole mass range are expected to decay to tau pairs with very significant branching ratios. The measurement of the Z production cross section described in Chap. 8 shows very good agreement with the theory and provides no clue of excess in the di-tau final states. However since the major background to the $H \to \tau\tau$ signal is $Z \to \tau\tau$ and the Higgs production cross section is expected to be very small, the presence of a Higgs signal could not be established in a general di-tau measurement but more complicated topologies could give increased sensitivity. For example the vector boson fusion(VBF) production of SM Higgs with tagging of the forward jets is expected to be more sensitive since the VBF signature requirement suppresses the Z background. In scenarios beyond the SM such as in the Minimal Supersymmetric Standard Model, the associated production of Higgs with b quarks allows the possibility to search for topologies with b-tagged jets to increase sensitivity. This chapter describes a search for SM and beyond the SM Higgs bosons in the di-tau final state.

9.1 Event Selection and Categorization

Both SM and MSSM searches are performed on a preselected data sample of di-tau events with specific additional requirements applied to enhance the sensitivity. In the case of the SM, the major production mechanisms are gluon fusion and vector boson fusion(VBF). The VBF production cross section is about ten times smaller than the gluon fusion production but the distinctive VBF signature enhances sensitivity by suppressing $Z \to \tau\tau$ background. In the case of gluon fusion, the Z background is overwhelming but the number of expected Higgs events is large. If the systematic uncertainties are small, gluon fusion can further improve the sensitivity achieved by VBF selection. The SM search is performed in the mass region between 110 and 145 Gev where the branching ratio of the Higgs boson to tau pairs is significant. In

M. Bachtis, *Heavy Neutral Particle Decays to Tau Pairs*,
Springer Theses, DOI: 10.1007/978-3-319-03257-3_9,

the case of the MSSM search, the main production mechanisms are gluon fusion and associated production with b-quarks. In the case of associated production, requiring the presence of b-jets in the final state suppresses Z background. The relevant mass region in the MSSM corresponds to a large mass window between 90 and 500 Gev since the main target particle is the pseudo-scalar A boson that can have very high mass.

The preselection of di-tau events is almost identical to the one used to measure the $Z \to \tau\tau$ production cross section in Chap. 8. Additional study is needed for the case when the kinematic requirements used in $Z \to \tau\tau$ analysis are inappropriate for the Higgs searches. In the case of the SM, the Higgs masses studied are low, resulting in very similar topologies as for the Z boson. On the other hand, for the MSSM Higgs search where the pseudo-scalar Higgs could be heavy, the transverse mass requirement applied in $\mu + \tau_h$ and $e + \tau_h$ final states shows inefficiency as the A mass increases. To overcome this limitation, the $M_T < 40$ Gev requirement is replaced by a collinearity requirement on the ζ variables given by:

$$p_\zeta^{\text{cut}} = P_\zeta - \alpha \cdot P_\zeta^{\text{vis}} > -20 \text{ Gev}, \tag{9.1}$$

with $\alpha = 1.2$ as introduced in Chap. 7. Figure 9.1 shows the product of acceptance times efficiency for the two kinematic selections. The p_ζ^{cut} requirement shows higher signal efficiency for higher Higgs masses, and is preferred for the MSSM search.

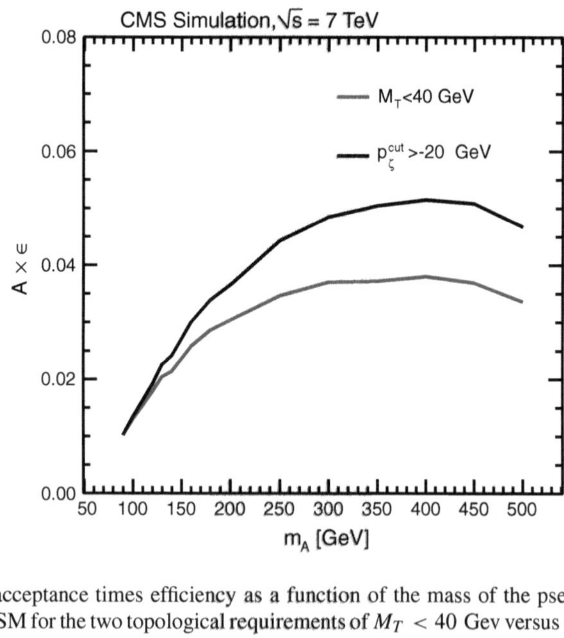

Fig. 9.1 Total acceptance times efficiency as a function of the mass of the pseudo-scalar Higgs boson in the MSSM for the two topological requirements of $M_T < 40$ Gev versus $p_\zeta^{\text{cut}} > -20$ Gev. For high masses of the boson, the collinearity variable p_ζ^{cut} provides higher signal efficiency

After basic pre-selection, the events are categorized in exclusive samples based on the specific requirements imposed to enhance the sensitivity. In the case of the Standard Model, a VBF category is introduced. The VBF signature is characterized by two jets with large pseudorapidity difference($\Delta\eta$) that due to the specific kinematic properties tend to have large invariant mass(M_{jj}). In addition, due to the absence of color connection between the two outgoing quarks, low QCD activity is expected in the pseudorapidity gap between the two jets. Figure 9.2 shows the number of jets with $p_T > 30$ Gev and $|\eta| < 4.5$, the pseudorapidity difference between the two tagged jets, the mass of the two tagged jets and the number of jets with $p_T > 30$ Gev in the pseudorapidity gap. Distributions are shown for $Z \to \tau\tau$ events and a Higgs sample of $m_H = 120$ Gev produced via gluon fusion and vector boson fusion. The VBF signature shows good separation in the studied variables from Z production.

For the SM Higgs search, the events are categorized as following:

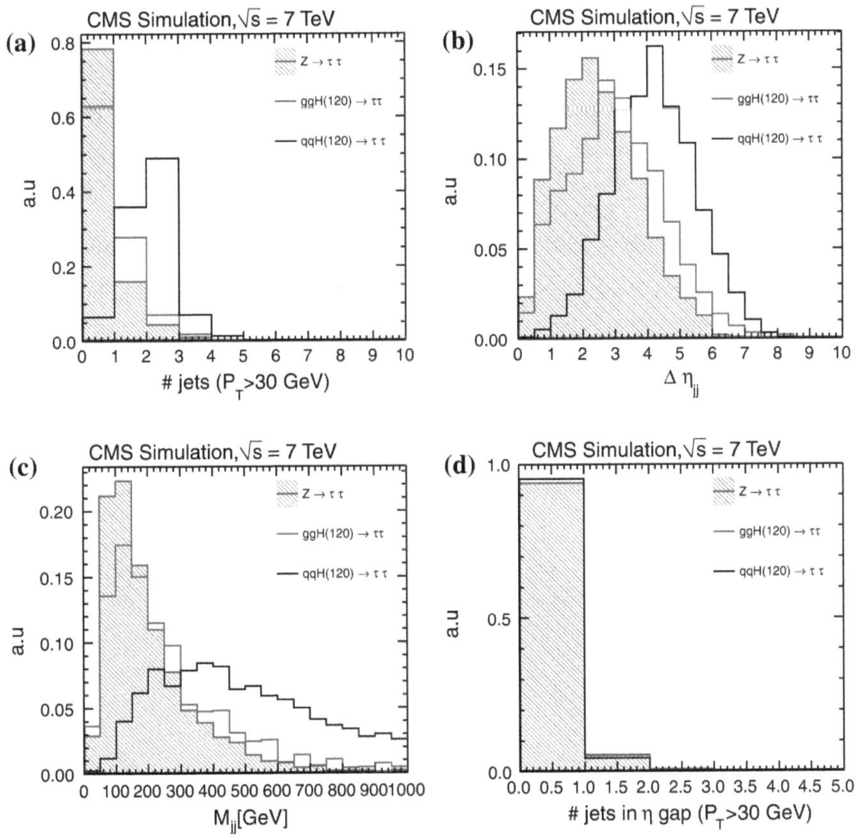

Fig. 9.2 VBF discrimination variables for Higgs and Z simulated samples, **a** Number Of jets with $p_T > 30$ Gev, **b** $\Delta\eta$ between tagged jets, **c** Mass of the tagged jet pair **d** Number of jets in the pseudorapidity gap between the tagged jets

- SM-VBF: Requires the presence of at least two jets with $p_T > 30$ Gev and $|\eta| < 4.5$. The jets must be outside a cone of $\Delta R = 0.5$ around the tau visible decay products and must have a pseudorapidity difference of $\Delta\eta > 4.0$. For the invariant mass of the two tagged jets, it is required to that $M_{jj} > 400$ Gev. In addition no additional jets are required in the gap between the two tagged jets with $p_T > 30$ Gev.
- SM-NoVBF: Contains all the events that pass the preselection but are not included in the VBF category.

The SM-VBF category is dominated by Vector Boson Fusion production while the SM-NoVBF category is dominated by gluon fusion. Both categories are combined in the final fit resulting in a combination of three final states with two categories each to estimate the Higgs yield.

In the case of the MSSM, the associated production with b quarks is exploited to enhance the sensitivity. Jets originating from b quarks are identified using the Track Counting High Efficiency algorithm discussed in Sect. 5.7. The jets are required to be within the tracker acceptance($|\eta| < 2.4$) and have a transverse momentum of $p_T > 20$ Gev. The events are categorized as following in the MSSM search:

- MSSM-B: Requires at least one b-tagged jet using the Track Counting High Efficiency (TCHE) algorithm (Sect. 5.7) with medium threshold. The jet must have $p_T > 20$ Gev and $|\eta| < 2.4$. In addition, events are vetoed if they have two or more jets with $p_T > 30$ Gev.
- MSSM-NoB: Contains all the events not present in the MSSM-B category.

The veto on the second jet above 30 Gev is effective against $t\bar{t}$ production. In the case of $t\bar{t}$, there are many energetic jets in the final state while in the case of heavy MSSM Higgs production, the second b jet tends to be soft and often not identified. Figure 9.3 shows the number of jets with $p_T > 30$ Gev and the number of TCHE b−tagged jets

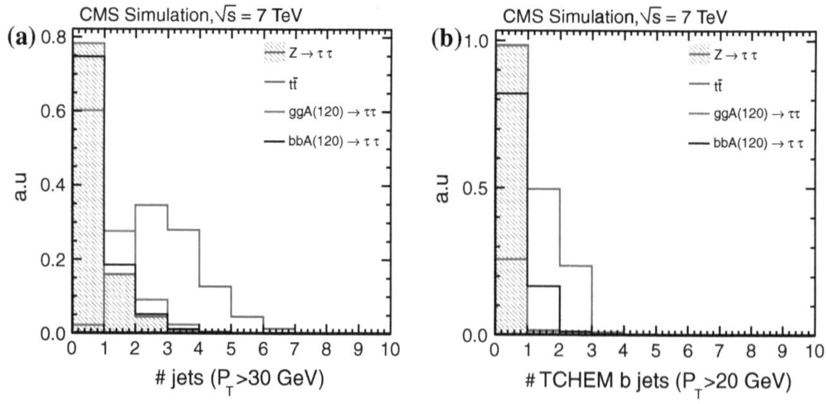

Fig. 9.3 MSSM Higgs discriminator variables for Higgs signal, $Z \to \tau\tau$ and $t\bar{t}$ samples, **a** Number of Jets with $p_T > 30$ Gev, **b** Number of identified TCHE Medium b jets with $p_T > 20$ Gev

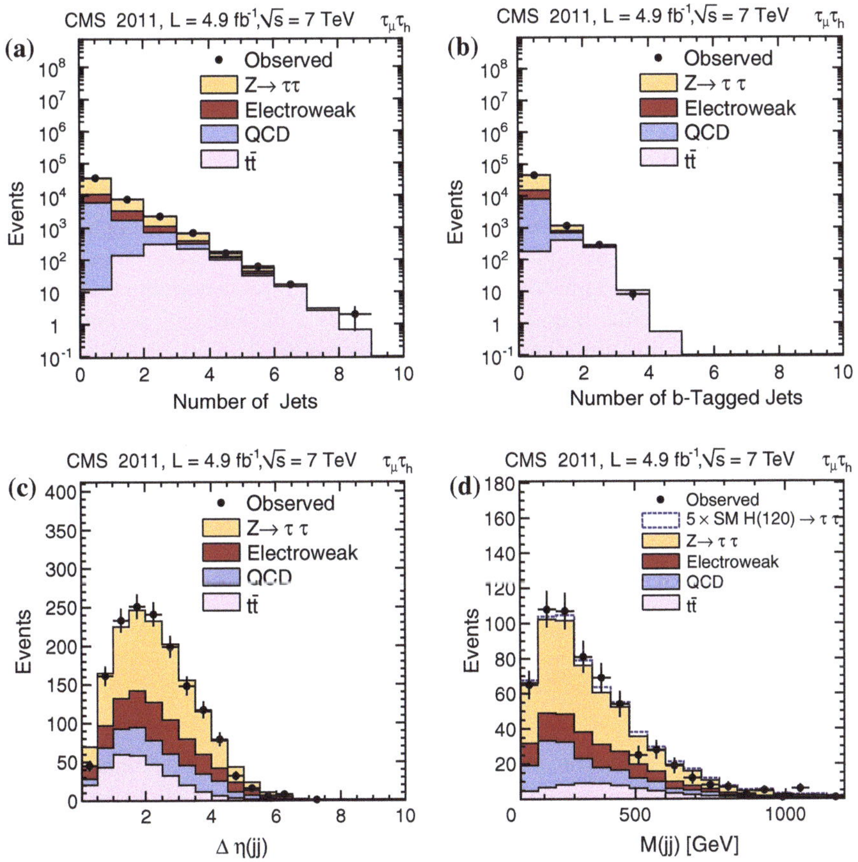

Fig. 9.4 Comparison of data and simulation for the main discrimination variables used to categorize the events in the Higgs search, **a** Number of Jets ($p_T > 30$ Gev), **b** Number of TCHEM b-jets, **c** $\Delta\eta$ between tagged jets, **d** M_{jj} between tagged jets

with medium threshold and $p_T > 20$ Gev. In the case of Z production and Higgs via gluon fusion, no b-tagged jets are expected while in the case of $bbA \rightarrow bb\tau\tau$ production, a second b-tagged jet is rarely identified. On the other hand, in the case of $t\bar{t}$ there is very often a second b-tagged jet and a lot of jets are reconstructed in the final state. Figure 9.4 shows comparison of the basic categorization variables between data and simulation. Good agreement is observed for all variables.

9.2 Background Estimation

The MSSM-NoB and SM-NoVBF categories consist of a huge number of events therefore the background estimation method is identical to the one used in the $Z \rightarrow \tau\tau$ cross section (Chap. 8). For the estimation of the background in the low statistics categories (MSSM-B, SM-VBF) specific procedures are used discussed in the following paragraphs. For all final states, the $Z \rightarrow \tau\tau$ and $t\bar{t}$ background are estimated using a common methodology. To estimate the Z contribution in the low-statistics categories, the embedded sample used in Sect. 7.3 is exploited. $Z \rightarrow \mu\mu$ events are selected with an associated b-jet or VBF signature and the muons are replaced with taus that are allowed to decay. The embedding procedure gives the shape of the Z background in the MSSM-B and SM-VBF categories and also an extrapolation factor that is used to give the Z yields. The extrapolation factor is defined as:

$$f_Z^B = \frac{N_{\text{emb}}^B}{N_{\text{emb}}}, \tag{9.2}$$

where N_{emb}^B is the number of embedded events that satisfy the requirements of the MSSM-B category and N_{emb} is the total number of embedded events. After calculating the extrapolation factor, the total number of Z events in the MSSM-B and SM-VBF categories is given by $N_Z^{ls} = f_Z^{ls} \cdot N_Z$ where N_Z is the number of Z events in the preselection that is given by simultaneous fit of the $Z \rightarrow \mu\mu$ sample described in Sect. 9.3

In the case of $t\bar{t}$ production, the simulation is used to estimate the number of final events. The simulation is normalized to the measured $t\bar{t}$ cross section. In the case of MSSM-B category where real $t\bar{t}$ events are present, the final yield is corrected for the difference of the b-tagging efficiency in data and simulation estimated by the CMS b-tagging group [1]. The QCD and fake background estimation is different depending on the studied final state.

9.2.1 QCD and W Background Estimation

For the estimation of the W background in the SM-NoVBF and MSSM-NoB categories, the same principle discussed in Chap. 7 is applied. In the case of the MSSM search, the p_ζ^{cut} requirement is inverted to define the W enriched region by requiring $p_\zeta^{\text{cut}} < -40$ Gev. The W background in the MSSM-B and SM-VBF categories is estimated by defining an extrapolation factor from the preselected estimated W events to the events in the final sample. This factor is taken from MADGRAPH simulation and a correction is applied based on the difference on the extrapolation factor for Z defined by using MADGRAPH Z+jets versus embedded sample. An additional systematic uncertainty of 10 % is applied. The QCD background needs to be estimated from the number of QCD events estimated in the preselected sample multiplied by an extrapolation factor to estimate the number of QCD events in the MSSM-B and SM-VBF categories. The extrapolation factor is defined in the signal free region defined by the Same Sign(SS) events. The SS sample is enriched in QCD by relaxing the lepton isolation to the isolation applied in the trigger and using only an additional requirement of

$$I_l^{PF}(\text{charged}) = \frac{\sum p_{\mathrm{T}}^{\text{charged}}}{p_{\mathrm{T}}^{l}} < 0.2 \tag{9.3}$$

In addition the tau identification is relaxed to satisfy only the Very Loose isolation discriminator defined in Chap. 6. The sample is dominated by QCD events and some residual W contribution that is subtracted using the M_T/p_ζ^{cut} sideband extrapolation method applied in Chap. 7. In this QCD enriched region, the MSSM-B or SM-VBF requirement is applied leading in a sample dominated by QCD and $t\bar{t}$ events. The $t\bar{t}$ events are subtracted using the estimation from simulation. Then, the QCD extrapolation factor is defined as:

$$f_{\text{QCD}}^{ls} = \frac{N_{ls}^{SS} - W_{ls}^{SS} - t\bar{t}_{ls}^{SS}}{N^{SS} - W^{SS} - t\bar{t}^{SS}} \tag{9.4}$$

where N^{SS}, W^{SS} and $t\bar{t}^{SS}$ is the number of data, estimated W and $t\bar{t}$ events in the SS preselected region and N_{ls}^{SS}, W_{ls}^{SS} and $t\bar{t}_{ls}^{SS}$ is the number of data, estimated W and $t\bar{t}$ events in the SS region that satisfies the low statistics (SM-VBF or MSSM-B) category requirements. The factor f_{QCD}^{B} is multiplied by the number of estimated QCD events in the preselected OS region to give the number of OS QCD events in the low statistics category.

9.2.1.1 Fake Background Estimation in the $e + \mu$ Final State

The fake background in the case of the $e + \mu$ final state is estimated using the ABCD method described in Sect. 7.6. The regions are modified to implement the categories as following:

- A: OS, both leptons isolated and MSSM-B or SM-VBF criteria applied
- B: OS, both leptons anti-isolated
- C: SS, both leptons isolated and MSSM-B or SM-VBF criteria applied
- D: SS, both leptons anti-isolated.

Tables 9.1, 9.2, 9.3 show the yields for all background samples after all efficiency corrections and data driven background estimation applied.

Figure 9.5 shows the visible mass for the MSSM-NoB and MSSM-B categories for all final states while Fig. 9.6 shows the visible mass for SM-NoVBF and SM-VBF categories for all final states after full background estimation applied. Good agreement is observed between data and simulation showing no significance excess of a signal.

Table 9.1 Event yields for all backgrounds after all corrections applied and after full background estimation for the $\mu + \tau_h$ final state for all event categories

	Standard model		MSSM	
Process	*No VBF*	*VBF*	*No B*	*B*
Di-Boson	219 ± 68	0.50 ± 0.50	274 ± 85	5 ± 3
$t\bar{t}$	834 ± 87	4 ± 1	1387 ± 155	159 ± 19
Z^{l+jet}	470 ± 515	1.5 ± 0.5	630 ± 68	3 ± 1
Z^{ll}	351 ± 107		406 ± 123	
$W + jets$	6012 ± 445	12 ± 5	10205 ± 744	115 ± 21
QCD	8262 ± 776	18 ± 3	6633 ± 689	187 ± 20
$Z \rightarrow \tau\tau$	30163 ± 2320	37 ± 4	31030 ± 2386	261 ± 25
Total Background	46311 ± 2493	73 ± 7	50565 ± 2603	730 ± 42
Observed	46389	74	50532	681

Table 9.2 Event yields for all backgrounds after all corrections applied and after full background estimation for the $e + \tau_h$ final state for all event categories

	Standard model		MSSM	
Process	*No VBF*	*VBF*	*No B-Tag*	*B-Tag*
Di-Boson	132 ± 41	0.15 ± 0.15	133 ± 41	2 ± 1
$t\bar{t}$	704 ± 74	4 ± 1	660 ± 74	76 ± 9
Z^{l+jet}	1968 ± 209	12 ± 2	1971 ± 209	40 ± 7
Z^{ll}	4261 ± 449		4241 ± 447	
$W + jets$	5999 ± 619	10 ± 5	5947 ± 614	51 ± 13
QCD	8576 ± 1224	8 ± 2	8515 ± 1216	119 ± 18
$Z \rightarrow \tau\tau$	14580 ± 1123	15 ± 2	14473 ± 1115	126 ± 13
Total Background	36221 ± 1842	51 ± 6	35939 ± 1829	415 ± 28
Observed	35479	44	35189	398

Table 9.3 Event yields for all backgrounds after all corrections applied and after full background estimation for the $e + \mu$ final state for all event categories

	Standard model		MSSM	
Process	*No VBF*	*VBF*	*No B-Tag*	*B-Tag*
Di-Boson	460 ± 139	0.8 ± 0.3	484 ± 146	5 ± 2
$t\bar{t}$	2688 ± 217	14 ± 2	2463 ± 199	360 ± 47
QCD/V+jets	1589 ± 181	7 ± 1	1560 ± 179	44 ± 6
$Z \rightarrow \tau\tau$	13908 ± 579	18 ± 1	14290 ± 930	154 ± 13
Total Background	18646 ± 659	40 ± 3	18797 ± 978	564 ± 49
Observed	18521	29	18092	519

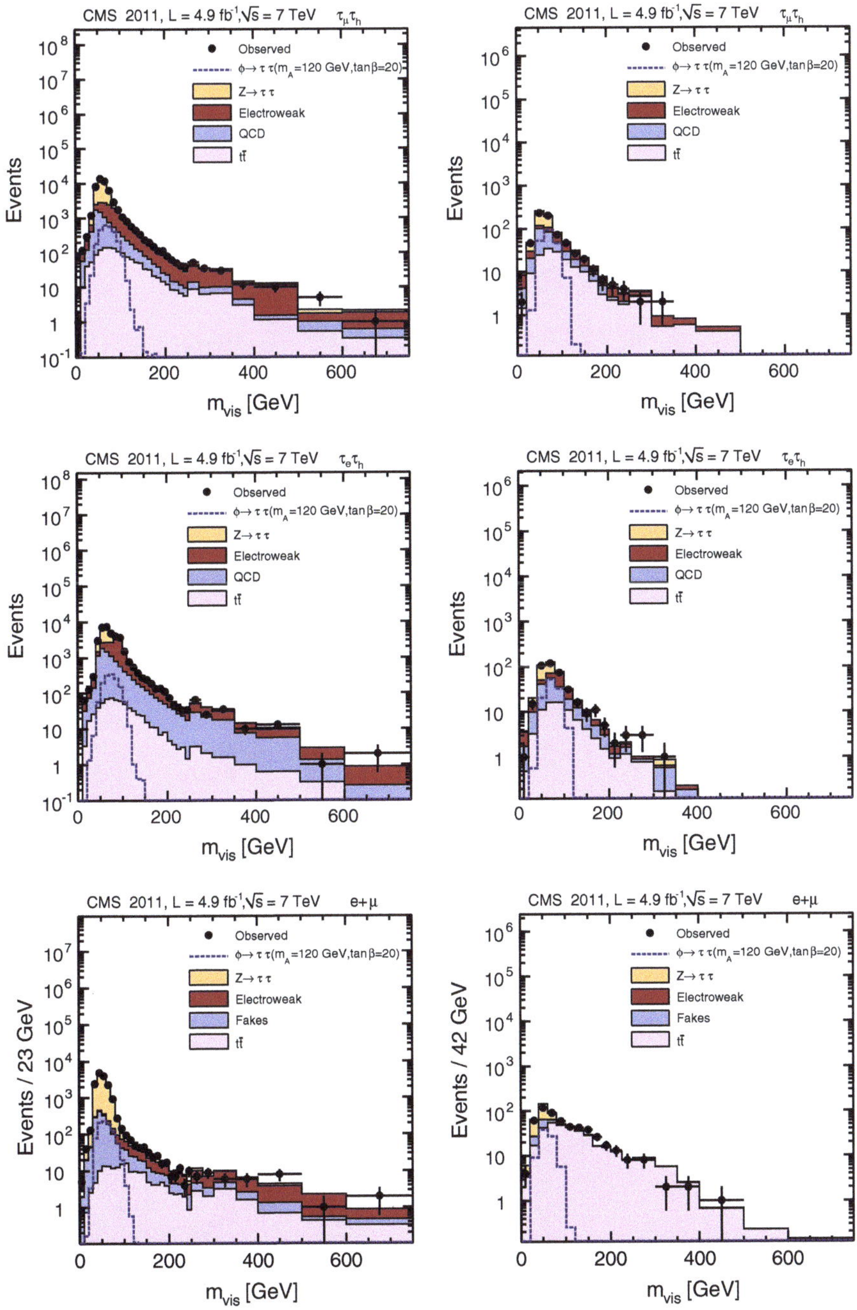

Fig. 9.5 Visible mass for the MSSM-NoB (*left*) and MSSM-B categories (*right*) for $\mu + \tau_h$ (*up*) $e + \tau_h$ (*center*) and $e + \mu$ final states (*down*)

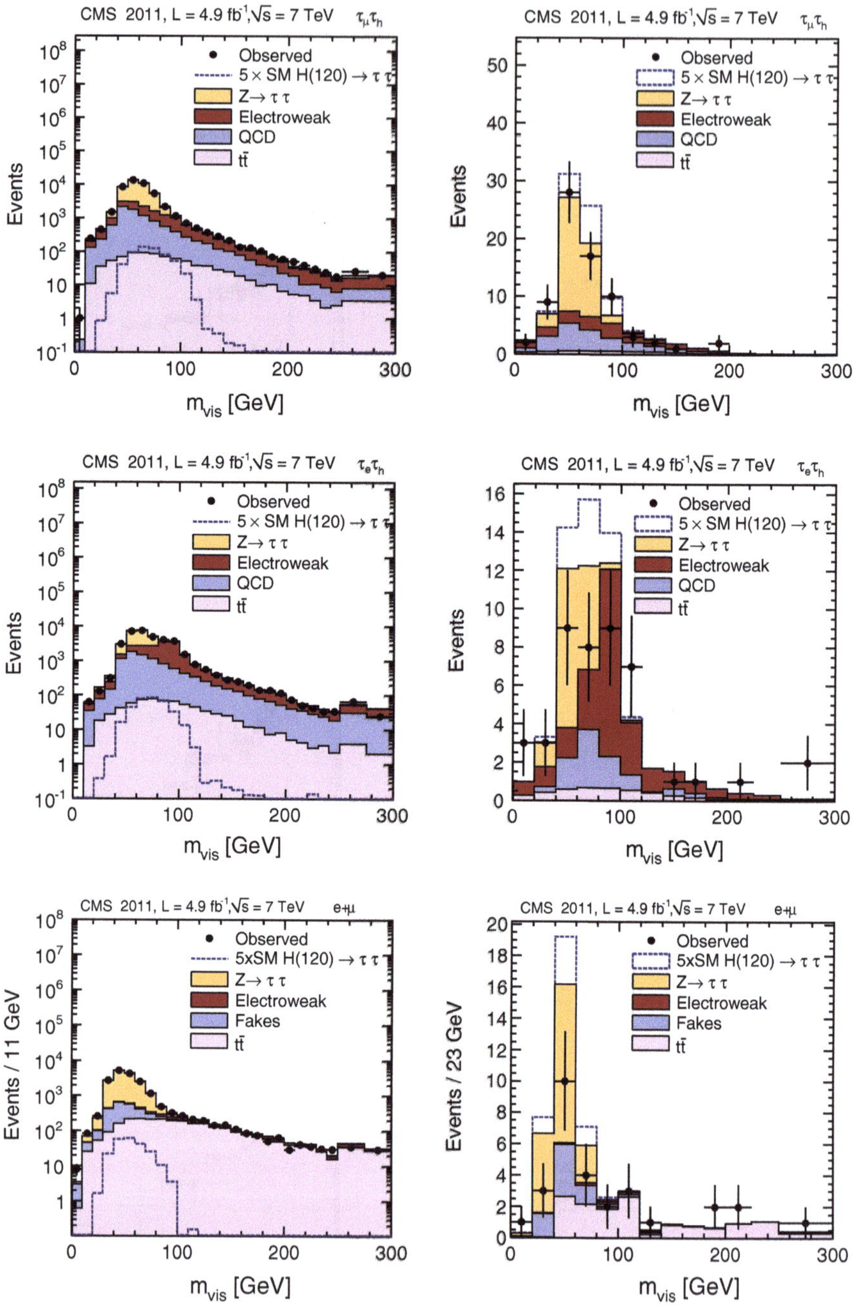

Fig. 9.6 Visible mass for the SM-VBF (*left*) and SM-VBF categories (*right*) for $\mu + \tau_h$ (*up*) $e + \tau_h$ (*center*) and $e + \mu$ final states (*down*)

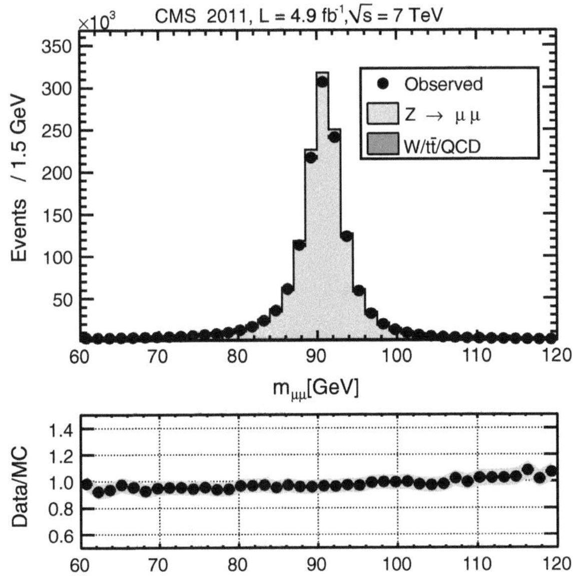

Fig. 9.7 Di-muon invariant mass for the selected Z enriched sample

9.3 Normalization of $Z \rightarrow \tau\tau$ Background for Higgs Searches

Since the major background to Higgs searches is $Z \rightarrow \tau\tau$ production, precise knowledge of the Z yield is essential to improve the Higgs sensitivity. The background cannot be normalized to the measured cross section discussed in Chap. 8 since the same sample is used to estimate the cross section and search for the Higgs and the Z fit does not account for presence of Higgs events in the sample. One choice would be to normalize the Z yield based on the theoretical cross section [2] using the integrated luminosity collected. The limitations to this approach are the theoretical uncertainty of 4.3 % from FEWZ calculation and the uncertainty on the integrated luminosity of the order of 2.2 %. To overcome those obstacles the Z normalization is estimated using collected $Z \rightarrow \mu\mu$ data. A sample is selected using di-muon triggers. Two muons are required with leading muon $p_T > 20$ Gev and sub-leading muon $p_T > 15$ Gev satisfying the criteria discussed in Sect. 5.3. The two legs are required to have opposite charge. Particle based isolation is used requiring $I_\ell^{PF/\Delta\beta} < 0.2$ for both legs (Sect. 5.5). Finally a mass window is selected of $60 < M_{\mu\mu} < 120$ Gev. The sample is dominated by $Z \rightarrow \mu\mu$ events while small backgrounds include W+jets, $t\bar{t}$ and QCD multijet events. Figure 9.7 shows the di-muon invariant mass after all selection requirements. A fit is performed in the sample to extract the Z correction factor. The likelihood of this final state is the same as the di-tau ones except the Higgs signal is negligible (primarily coming through $Z \rightarrow \tau\tau \rightarrow \mu\mu$ events). The small backgrounds are normalized using their expected yields and assigning a conservative uncertainty of 30 %. This way the Z correction factor is estimated during the fit and without any dependence in the luminosity uncertainty resulting in much higher precision for the $Z \to \tau\tau$ background. The result of the fit yields a correction factor for the MADGRAPH MC equal to

$$\rho_Z = 0.99 \pm 0.03 \tag{9.5}$$

9.4 Systematic Uncertainties

The dominant sources of experimental systematic uncertainties are identical to the ones described in the $Z \to \tau\tau$ cross section measurement (8.2.1) and correspond to the muon, electron and tau ID efficiency,energy scale and the uncertainties due to the embedding samples. For the MSSM categories, an additional uncertainty that is introduced to account for the uncertainty of the b-tagging efficiency and mis-identification rate. The b-tagging efficiency correction factor is estimated in $t\bar{t}$ events to be 0.96 ± 0.04 while the b mis-identification rate correction factor is estimated to be 1.20 ± 0.15 [1]. The correction factors affect only the signal and $t\bar{t}$ since in the case of other backgrounds the estimation of the background is performed from data. In the case of the MSSM signal, since there are cases where the b-tagged jet is real or fake, the corrections are applied in the simulation after matching the b-tagged jet to the generator level parton to define its flavor. For each event, if the b-tagged jet is b the event is weighted by the b-tagging efficiency correction otherwise it is weighted by the b mis-identification rate correction. In the same manner, in the sample with no b-tag if there is a real b jet and it is not b-tagged, the event is weighted by a correction factor corresponding to $1-$ efficiency. With this separation of the signal samples at the generator level, the correlations between the categories concerning the b-tagging efficiency and mis-identification rate are treated properly. Another relevant experimental source of uncertainty is the jet energy scale. The MSSM-B category requires a b-jet above 20 Gev, and maximum one jet of $p_T > 30$ Gev, therefore differences in the jet energy scale could allow migration between the categories or events that migrate outside the samples. Similarly in the SM-VBF category, variations of the energy scale could result in migration of events between the categories. This is relevant for the signal and for backgrounds that are estimated from simulation while in the case of $Z \to \tau\tau$ and data driven backgrounds the yields are estimated from data. The jet energy scale uncertainty varies as a function of p_T and η therefore to evaluate the effect on each yield, the jet transverse energy is varied by one standard deviation and the difference in the yield after all selection requirements is quoted as the systematic uncertainty. This systematic uncertainty is considered anti-correlated between the MSSM-B and MSSM-NoB categories and between SM-VBF and SM-NoVBF categories respectively. The experimental systematic uncertainties are summarized in Table 9.4. In terms of theoretical uncertainties, the dominant contributions are related to the uncertainties in the theoretically predicted Higgs cross section in the Standard Model or the MSSM. Different contributions to the cross section uncertainties are

Table 9.4 Summary of the experimental systematic uncertainties

Source	$\mu + \tau_h$%	$e + \tau_h$%	$e + \mu$%	SM	MSSM
Muon identification and trigger	1 %	–	2 %	√	√
Electron identification and trigger	–	1 %	2 %	√	√
Tau identification	6 %	6 %	–	√	√
Tau trigger	3.3 %	3.3 %	–	√	√
Efficiency of topological selections	0.5 %	0.5 %	0.5 %	√	√
Tau energy scale uncertainty	3 %	3 %	–	√	√
Jet energy scale uncertainty	2 – 5 %	2 – 5 %	2 – 5 %	√	√
$\not{E}_T$ scale uncertainty	5 %	5 %	5 %	√	√
b-tagging efficiency	4 %	4 %	4 %		√
b-tagging mis-ID rate	15 %	15 %	15 %		√
Luminosity	2.2 %	2.2 %	2.2 %	√	√

Table 9.5 Summary of the theoretical systematic uncertainties

	MSSM		SM	
Source	NoB%	B%	NoVBF%	VBF%
Parton distribution functions (qqH)	1 %	–	8 %	8 %
Parton distribution functions (ggH)	2 %	2 %	8 %	8 %
Parton distribution functions (bbH)	3 %	3 %	–	–
Factorization/renormalization scale (ggH)	13 %	13 %	12 %	30 %
Factorization/renormalization scale (qqH)	–	–	4 %	4 %
Factorization/renormalization scale (bbH)	20 %	20 %	–	–
Underlying event/parton shower	–	–	4 %	4 %

considered as separate uncertainties. The major uncertainties are the uncertainties due to using different PDF sets or taking into account the uncertainties of each PDF set. Additional, uncertainties are derived by varying the renormalization and factorization scale by 100 % or are due to contributions of the Underlying event and Parton shower simulation. The theoretical systematic uncertainties are different for each production mechanism and are included in Table 9.5.

9.5 Combined Fit

In the case of searches for new physics, the likelihood model is very similar to the one described in Chap. 8, utilizing a likelihood that contains shape, background and systematic effects contributions. The shape background in the case of searches is modified as follows:

$$\mathcal{L}_{\text{shape}} = r \times \sum A_k(n_j) \cdot f^k_{H\to\tau\tau}(m_{\text{vis}}, n_j) + \sum N_i(n_j) f_i(m_{\text{vis}}, n_j), \quad (9.6)$$

The $Z \rightarrow \tau\tau$ contribution is now counted as a background and the first sum over k corresponds to the signal contributions that can be as many as the possible separate Higgs production mechanisms. The parameters r and A_k are selected on the basis of the quantity to be estimated and are discussed in the following paragraphs. For each specific analysis, the likelihood is minimized in a similar way to the $Z \rightarrow \tau\tau$ measurement and the treatment of the shape uncertainties is identical.

9.5.1 Definition of the SM Signal Model

In the case of the Standard Model Higgs search, the dominant production mechanisms are gluon fusion and vector boson fusion. Additional production mechanisms correspond to Higgsstralung(VH) and associated production with top pairs(ttH). In the Higgsstralung process, Higgs is produced in association with a vector boson while in the ttH process, Higgs is produced in association with a top quark pair. Three signal templates are used for the SM Higgs signal, one corresponding to gluon fusion, one corresponding to VBF and one corresponding to VH and ttH. Each template is normalized to the expected Higgs yield when the Higgs cross section is equal to the SM Higgs cross section. The cross section for each process and branching ratios are taken from the LHC Higgs cross section working group [3]. The extracted value of the combined fit is the signal strength defined as:

$$r = \frac{\sigma_{obs}}{\sigma_{SM}}, \tag{9.7}$$

where σ_{obs} is the observed Higgs cross section times branching ratio to tau pairs and σ_{SM} is the theoretical SM Higgs production cross section times branching ratio of Higgs to tau pairs.

9.5.2 Definition of the MSSM Signal Model

Since in the case of the MSSM there are three types of neutral Higgs bosons produced by two production mechanisms (gluon fusion and associated production) that have different magnitudes in the $m_A - \tan\beta$ plane, a specific model of the signal is required. In this analysis the results are interpreted against the $m_h^{\max}$ scenario [4]. The production cross section and the mass for each of the three bosons (A, h, H) is known at each specific value in the parameter space. For masses of the pseudo-scalar Higgs boson of $m_A < 130$ Gev the h and A bosons are degenerate while H boson is fixed at about 130 Gev . For masses of $m_A > 130$ Gev, A and H are degenerate while h is fixed at about 130 Gev. For a complete description of the Higgs signal, four templates are used:

- A template of the sum of the two degenerate Higgs bosons at the value of m_A for the gluon fusion production mechanism. The relative contributions of the two bosons are scaled to their production cross section for the specific value in the MSSM parameter space.
- A template of the sum of the two degenerate Higgs bosons at the value of m_A for the associated production with b-quarks mechanism. The relative contributions of the two bosons are scaled to their production cross section for the specific value in the MSSM parameter space.
- A template of the fixed Higgs boson at the mass of 130 Gev for the gluon fusion production mechanism.
- A template of the fixed Higgs boson at the mass of 130 Gev for the associated production with b-quarks mechanism.

Each template in the categories described above is given the specific Higgs bosons mass at the MSSM parameter space point by using linear interpolation between mass templates.

9.5.3 Statistical Analysis

The statistical methodology used in this analysis has been developed by the CMS and ATLAS collaborations in the context of the LHC Higgs Combination group. The detailed description of the statistical procedure can be found elsewhere [5]. A parameter of interest r is defined that corresponds to the signal strength in the case of the SM Higgs boson search or the $\tan\beta$ in the case of the MSSM Higgs search. Each systematic uncertainty is assigned a nuisance parameter θ_i. The expected signal and background yields are denoted as $r \cdot s(\theta)$ and $b(\theta)$ respectively. Most nuisance parameters are constrained by other measurements or theoretical predictions and are encoded in the probability density functions $p_i(\theta\,|\theta)$, corresponding to the probability to measure a value θ_i for the i−th nuisance parameter given its real value θ_i. The Likelihood is given by

$$\mathcal{L}(\text{data}|r \cdot s(\theta) + b(\theta)) = \mathcal{P}(\text{data}|r \cdot s(\theta) + b(\theta)) \cdot p(\theta\,|\theta), \tag{9.8}$$

where $\mathcal{P}(\text{data}|r \cdot s(\theta) + b(\theta))$ is a product of the probabilities over all shapes in all channels and $p(\theta\,|\theta)$ is the product of all the nuisance constraining probability density functions containing the nuisance and background terms $\mathcal{L}_{\text{syst}}$ and $\mathcal{L}_{\text{bkg}}$. In order to test the signal hypothesis, an appropriate test statistic is created. A test statistic is a single number that holds information on the observed data, expected signal, background and uncertainties. To test the absence of a signal the test statistic value estimated using the data is compared to the distribution of the test statistic under the signal and background vs the background only hypothesis. The expected test statistic distributions are calculated by generating pseudosamples from the probability density functions $\mathcal{P}(\text{data}|r \cdot s(\theta) + b(\theta))$ and $p(\theta\,|\theta)$. The values of nuisance parameters θ used for generating pseudo-datasets are obtained by maximizing the likelihood $\mathcal{L}$ under the background only or signal+background hypothesis.

To quantify the absence of a signal and set exclusion limits, a test statistic q_r is defined that depends on the hypothesized value of the parameter of interest r defined as:

$$q_\mu = -2\ln\frac{\mathcal{L}(\text{data}|r\cdot s(\hat{\theta}_r)+b(\hat{\theta}_r))}{\mathcal{L}(\text{data}|\hat{r}\cdot s(\hat{\theta})+b(\hat{\theta}))}, \tag{9.9}$$

where $\hat{\theta}$ and $\hat{r}$ are the values of θ and r that maximize the likelihood in numerator and denominator and θ_r implies that the minimization is done under the hypothesis of specific value of the parameter of interest r. One sided limits are forced by requiring that $r > \hat{r}$. For the calculation of the exclusion limit the modified frequentist approach of CL_s is used [6, 7]. The probability to obtain a value of the test statistic q_r larger than that the observed value q_r^{obs} is defined for the signal+background and background only hypothesis:

$$\text{CL}_{s+b} = \mathcal{P}(q_r \geq q_r^{\text{obs}}|r\cdot s+b) \tag{9.10}$$

$$\text{CL}_b = \mathcal{P}(q_r \geq q_r^{\text{obs}}|b). \tag{9.11}$$

CL_s is then given by the ratio:

$$\text{CL}_s = \frac{\text{CL}_{s+b}}{\text{CL}_b} \tag{9.12}$$

For the values of $\text{CL}_s \leq \alpha$ and $r = r_0$ the signal is excluded in the $1-\alpha$ confidence level for a value of the parameter of interest larger that $r = r_0$. To quote the upper limit on r at 95 % confidence level, the values of r are scanned . For each value of r distributions of the test statistic are created and the value of CL_s is calculated. The 95 % CL upper limit is set at the value of r where $\text{CL}_s = 0.05$. In the case of excess in the data, the significance of the excess is quantified by introduced an appropriate test statistic q_0 as follows:

$$q_\mu = -2\ln\frac{\mathcal{L}(\text{data}|b(\hat{\theta}_0))}{\mathcal{L}(\text{data}|\hat{r}\cdot s(\hat{\theta})+b(\hat{\theta}))}, \tag{9.13}$$

where θ_0 is the value of the nuisance parameters that maximizes the numerator under the background only hypothesis. The excess can be quantified in terms of the p-value p_0 defined as the probability to observe a value of the test statistic q_0 at least equal or larger than the one observed in data (q_0^{obs}) under the background only hypothesis:

$$p_0 = \mathcal{P}(q_0 \geq q_0^{\text{obs}}|b) \tag{9.14}$$

The p-value can then be translated to significance Z using the Gaussian one sided tail integral:

$$p_0 = \int_Z^\infty \frac{1}{\sqrt{2\pi}}\exp(-x^2/2)\text{d}x \tag{9.15}$$

9.6 SM Higgs Results

In the case of the SM search the NoVBF and VBF categories are combined in all three final states and a combined fit is performed to extract the signal strength for each Higgs mass hypothesis. No excess is observed in the visible mass spectra therefore upper limits are set to the Higgs signal strength. Limits are set using the modified frequentist CL_s construction. Distributions of the test statistic are created in a grid of (m_H, r) values to test the value of r where $CL_s = 0.05$ for each Higgs mass hypothesis.

Figure 9.8 and Table 9.6 show the expected and observed 95 % CL Upper Limits on the signal strength as a function of mass of the Higgs boson. This analysis is expected to exclude a Higgs boson of $m_H = 115$ Gev with a cross section of $\sigma > 3.2\sigma_{SM}$. The actual observation on the data varies from $r > 2.56$ to $r > 7.0$ with the best exclusion at $m_H = 115$ Gev with $\sigma > 2.56\sigma_{SM}$.

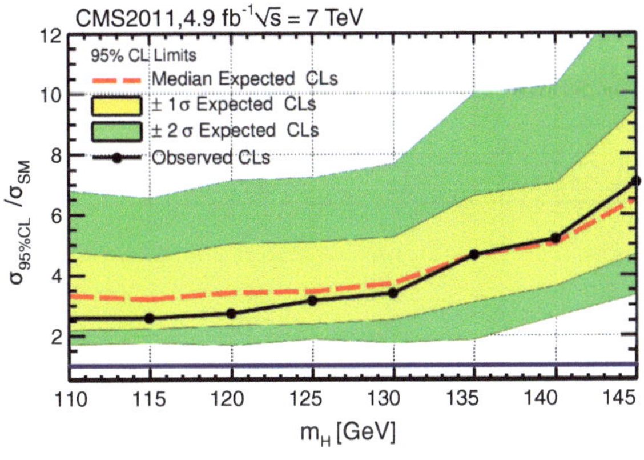

Fig. 9.8 Expected and Observed 95 % CL Upper Limits on the signal strength $r = \sigma_{obs}/\sigma_{SM}$ as a function of the SM Higgs boson mass for the SM Higgs search

Table 9.6 Expected and observed upper limits on the signal strength $r = \sigma_{obs}/\sigma_{SM}$ for the SM Higgs search

SM Higgs		Expected σ/σ_{SM} limit				
m_H [Gev]	Observed	−95 %	−68 %	Median	+68 %	+95 %
110	2.567	1.715	2.197	3.321	4.766	6.824
115	2.565	1.776	2.232	3.207	4.568	6.578
120	2.724	1.684	2.337	3.418	5.048	7.166
125	3.146	1.886	2.394	3.452	5.099	7.248
130	3.371	1.766	2.530	3.705	5.248	7.693
135	4.641	1.862	3.104	4.624	6.624	10.000
140	5.166	2.622	3.633	5.014	7.038	10.280
145	7.043	3.343	4.687	6.489	9.493	13.607

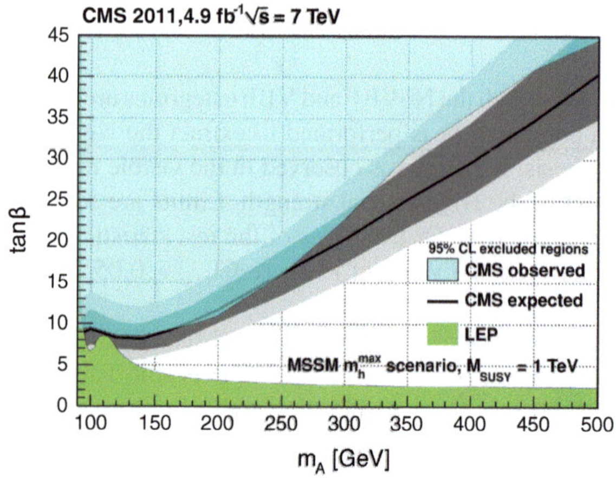

Fig. 9.9 Expected and Observed 95 % CL Upper Limits on the $\tan\beta$ as a function of the pseudoscalar A boson mass for the MSSM search

Table 9.7 Expected range and observed 95 % CL upper limits for $\tan\beta$ as a function of m_A, for the MSSM search

MSSM Higgs		Expected $\tan\beta$ limit				
m_A [GeV]	Obs. $\tan\beta$ limit	-2σ	-1σ	Median	$+1\sigma$	$+2\sigma$
90	9.114	5.477	6.721	8.541	10.257	12.619
100	9.577	6.089	7.473	9.307	11.764	13.893
120	8.510	5.396	6.941	8.307	10.324	12.503
130	8.494	6.084	6.893	8.239	10.030	12.220
140	8.426	5.690	6.776	8.172	10.065	12.173
160	9.174	6.412	7.266	9.081	11.102	13.063
180	9.919	7.297	8.692	10.327	12.281	14.654
200	10.861	8.202	10.149	11.702	14.182	16.558
250	15.927	12.469	13.471	16.145	19.061	20.000
300	22.697	15.082	17.459	20.315	23.773	27.435
350	30.440	19.520	21.788	25.188	29.286	33.000
400	35.649	22.491	25.782	29.641	34.529	39.975
450	40.904	26.265	30.880	34.774	41.000	47.181
500	44.481	31.361	34.924	40.389	47.177	54.869

9.7 MSSM Higgs Results

In the case of the MSSM, since the masses of the Higgs bosons, and the relative cross section from gluon fusion and associated production with b quarks change as a function of m_A and $\tan\beta$, the signal hypothesis is tested for each point in the parameter space. No significant excess is observed in the MSSM-NoB and MSSM-B categories therefore upper limits are set on $\tan\beta$ for each m_A hypothesis. To estimate the CL_s limits, distributions of the test statistic are generated in a grid of $(m_A, \tan\beta)$ values. For each point in the grid the signal templates of the h, H and A bosons are assigned their exact mass and the cross sections are constrained to the expected values for those parameters. Then for each m_A hypothesis the value of $\tan\beta$ with $CL_s = 0.05$ is defining the exclusion. Figure 9.9 and Table 9.7 show the expected and observed limits in the MSSM parameter space.The analysis is expected to exclude a value of $\tan\beta = 8.1$ for $m_A = 140$ Gev and a $\tan\beta = 40$ for $m_A = 500$ Gev. The actual exclusions in data at low mass is $\tan\beta < 8.4$ for $m_A = 140$ Gev and at high m_A of 500 Gev, the observed exclusion is $\tan\beta < 45$. The maximum excess appears at $(M_A, \tan\beta) = (350, 20)$ with a local significance of $Z = 1.53$ standard deviations.

References

1. The CMS Collaboration. b-jet identification in the cms experiment. CMS Physics Analysis Summary, (CMS-PAS-BTV-11-004), 2011
2. R. Gavin, Y. Li, F. Petriello, S. Quackenbush, FEWZ 2.0: a code for hadronic Z production at next-to-next-to-leading order (2010)
3. LHC Higgs Cross Section Working Group, S. Dittmaier, C. Mariotti, G. Passarino, R. Tanaka (Eds.), et al. Handbook of LHC Higgs Cross Sections: Inclusive Observables (2011)
4. M.S. Carena et al., MSSM Higgs boson searches at the Tevatron and the LHC: impact of different benchmark scenarios. Eur. Phys. J. **C45**, 797–814 (2006)
5. ATLAS, CMS Collaborations, and LHC Higgs Combination Group. Procedure for the LHC Higgs Boson Search Combination in Summer 2011. ATL-PHYS-PUB-2011-011, CMS NOTE-2011/005 (2011)
6. T. Junk, Confidence level computation for combining searches with small statistics. Nucl. Instrum. Meth. A **434**, 435–443 (1999)
7. A. L. Read. Modied frequentist analysis of search results (the CLs method). Yellow Report, CERN-2000-005:81–101 (2000)

Chapter 10
Synopsis

In the previous chapters, a complete study of the di-tau final state at $\sqrt{s} = 7\,\text{TeV}$ was presented using the complete dataset collected by the CMS Detector during the 2011 LHC run.

The key to this study is the capability to efficiently reconstruct taus that decay hadronically in the dense LHC enviroment with many QCD jets in the final state and many pileup events superimposed in each collision. To achieve this goal the Hadrons Plus Strips(HPS) algorithm which uses all reconstructed particle candidates to build the tau decay products was developed. HPS tau identification introduces for first time in hadronic collisions combinatorial reconstruction of the tau decay modes. The fake rates achieved for loose isolation are of the order of 1 %, maintaining 50 % of the real hadronically decaying tau leptons.

Using the HPS algorithm in combination with the excellent muon, electron and jet reconstruction, events are selected and analyzed in the three major di-tau final states. An important challenge in the 2011 run was the increase of the average number of pileup interactions, affecting mainly lepton and tau isolation in this analysis. For this purpose, pileup mitigation using $\Delta\beta$ corrections was introduced for the isolation of electrons, muons and taus maintaining the signal efficiency for small increase of the overall background rates.

Since, there is significant background contamination from QCD and W+jets after full event selection, background estimation techniques were designed, tested and used since the beginning of the CMS data taking to predict those backgrounds with high accuracy. Those methods exploit the W transverse mass shapes and the charge of the di-tau candidate giving accurate predicitions that are cross checked by the simulation and other approaches.

Using the techniques described above, the $Z \rightarrow \tau\tau$ cross section was measured with unprecedented accuracy to be in agreement with the theoretical expectations. This measurement constitutes a validation of the HPS tau identification and background estimation techniques. After establishing $Z \rightarrow \tau\tau$ and gaining confidence in all the procedures used, a search for the Higgs boson is performed in the Standard Model and the Minimal Supersymmetric Standard Model. Both searches show no

M. Bachtis, *Heavy Neutral Particle Decays to Tau Pairs*,
Springer Theses, DOI: 10.1007/978-3-319-03257-3_10,

evidence for Higgs signal. Stringent new bounds in the SM and MSSM searches in the di-tau final state are set. In the case of the Standard Model Higgs search the sensitivity is not yet adequate to probe the signal asuming the theoretical cross sections and an 95 % CL upper limit is set excluding a Higgs production cross section with value higher than about three times the theoretical one for low Higgs boson mass. In the case of the Minimal Supersymmetric Standard Model, stringent new bounds are set in the MSSM parameter space (m_A, $\tan\beta$) restricting the MSSM to lower values of $\tan\beta$ and higher masses of the pseudoscalar A boson.

The LHC is going to operate in 2012 with higher instantaneous luminosity and a center of energy of 8 TeV. Considering the new run parameters and possible analysis improvements, it is certain that with a sample of 20 fb^{-1} of 2012 data, this analysis will play a major role in discovering or excluding a Higgs boson at low mass.

Finally the future of this analysis was considered from the scope of the trigger needed to collect the di-tau data in a much higher instantaneous luminosity scenario with a lot of pileup. For this purpose new calorimeter trigger algorithms were proposed that can be implemented in commercial FPGAs and are the basis of a proposal for the CMS Calorimeter Trigger upgrade. Those algorithms provide much higher efficiency and lower rate for the tau trigger making possible an efficienct selection of the di-tau sample at L1 and maintaining particle thresholds that are already used in the analysis. The implemented calorimeter trigger algorithms are described in Appendix.

Appendix
Calorimeter Trigger Upgrade For Higher Luminosities

The $H/Z \to \tau\tau$ analysis presented in the previous chapters has been designed to be robust againt pileup. However a possible future limitation to the performance will be the capability to collect the events needed for the analysis at trigger level. In the 2011 run, events in the $\mu + \tau_h$ and $e + \tau_h$ final states were triggered using single electron or muon triggers at L1 and cross triggers at HLT. As the instantaneous luminosity increases, the L1 rate of the single object triggers will not be affordable, requiring the implementation of a lepton+tau trigger at L1. As discussed in Sect. 6.5, the main limitations at L1 are the measurement of the tau energy with the calorimeter, the limited granularity of the trigger towers and the possible isolation only using calorimeter deposits. However, a combined cross trigger at L1 with an isolated e/γ and a tau candidate with low threshold would improve the $e + \tau_h$ final states. In addition, the increased occupancy of each crossing degrades the performance of the implemented electron and tau trigger algorithms, which leads to the need of more sophisticated object identification in the L1 trigger. This chapter describes a proposed set of calorimeter trigger algorithms that provide improved performance and can be used as a basis of a calorimeter trigger upgrade in the following years to come. CMS is planning for an upgrade for the SLHC era where the LHC will be upgrated to provide instantaneous luminosities of the order of 5×10^{34} cm^{-2}s^{-1} resulting in about 100 pileup interactions per crossing at a 25 ns bunch crossing separation. However, since the required technology already exists, this upgrade could be performed sooner to boost the performance of the detector towards Higgs properties measurement or Higgs exclusion in the following years.

The requirements of the upgraded calorimeter trigger algorithm in comparison to the present calorimeter trigger can be summarized as follows:

- Provide equal or better performance for electromagnetic, τ and jet objects by keeping the rates low in the presence of pileup.
- Provide the best possible position resolution to improve cross trigger performance and to match calorimeter objects to tracks in case of a future tracker upgrade providing a track trigger at L1.

M. Bachtis, *Heavy Neutral Particle Decays to Tau Pairs*,
Springer Theses, DOI: 10.1007/978-3-319-03257-3,

- Exploit the latest technologies to create fast, flexible and reconfigurable hardware to be adaptable to any conditions.

As in the present calorimeter trigger, the energy deposit signatures the trigger aims at distinguishing are those from electrons or photons, hadronic τ lepton decays, and jets. The Moliere radius of the CMS Electromagnetic Calorimeter is well below the size of single trigger tower. Hence, in the absence of bremsstrahlung and material effects, electromagnetic deposits that hit the center of a trigger tower are contained within one tower. Taking bremsstrahlung, material and boundary effects into account, electrons and photons deposit their energy in clusters of at most $\Delta(\eta) \times \Delta(\phi) = 2 \times 2$ trigger towers. The case of hadronic τ decays is more complex due to range of possible decay products. However, a hadronic τ lepton decay typically leads to a well collimated high energy deposit in the calorimeters. In order to control the huge rate from QCD jets, the τ candidate energy needs to be measured in a small region. The optimal region is found to be $\Delta(\eta) \times \Delta(\phi) = 2 \times 2$ trigger towers as well. The largest performance gains at the SLHC are expected from a redesign of the electromagnetic and the τ trigger algorithms described in the following paragraphs.

1 Input/Output Specification

The studies presented in this thesis assume that the granularity of the calorimeter trigger towers at remains unchanged with respect to the present conditions in the ECAL and HCAL barrel and endcaps. The granularity is expected to increase in the HF. In the range $-3 < \eta < 3$, there is a two-fold division in depth, i.e. separate information for ECAL and HCAL is assumed. The expected trigger tower dimensions in $\Delta(\eta) \times \Delta(\phi)$ are summarized in Table 1 for positive η values. The binning is assumed symmetric in η. Particle occupancy from single non-diffractive interactions is roughly contant in η.

The input information per tower consists of 14 bits of non-linear E_T information (7 for ECAL and 7 for HCAL) accompanied by a single feature bit determined from a fine grain analysis of the electromagnetic energy profile within the tower.

Table 1 Trigger tower dimensions

η tower count	η range	$\Delta\eta$	ϕ tower count	$\Delta\phi$	ECAL	HCAL
1–21	0.00–1.83	0.087	72	0.087	EB or EE	HB or HE
22	1.83–1.93	0.100	72	0.087	EE	HE
23	1.93–2.04	0.113	72	0.087	EE	HE
24	2.04–2.17	0.129	72	0.087	EE	HE
25	2.17–2.32	0.150	72	0.087	EE	HE
26	2.32–2.50	0.178	72	0.087	EE	HE
27	2.50–2.65	0.150	72	0.087	EE	HE
28	2.65–3.00	0.350	72	0.087	EE	HE
29–40	3.00–5.00	0.167	36	0.175	-	HF

The proposed output consists of the highest E_T objects in three categories: 4 electromagnetic objects, 4τ objects and 12 jet objects, ranked by E_T; plus a set of global event characteristics: missing transverse energy (MET), total transverse energy (SumET), total jet transverse energy (HT) and missing jet transverse energy (MHT). For electromagnetic and τ objects, transverse energy information is the sum of 2×2 tower E_T corresponding to 10 bits. The position of the cluster can be approximated to better accuracy than one tower by weighting the cluster position. Therefore for the position of the cluster one bit is allocated for the side of the detector, five bits for the tower number and four bits for the internal cluster position, corresponding to 10 bits total.

2 Description of the Upgrade Algorithms

The proposed algorithms create collections of isolated and non-isolated electromagnetic objects, isolated and non-isolated τ objects and jet objects. The algorithm is organized in several steps with progressive data reduction. The functionality of the steps could be implemented each in separate types electronic cards.

1. **Particle Cluster Finder:** Reconstructs 2×2 clusters that overlap in η and ϕ directions by one trigger tower. Identifies if the cluster transverse energy profile is compatible with an electron or photon.
2. **Cluster Overlap Filter:** Identifies local maxima of energy and applies prunning of towers in the clusters that overlap with a neighboring cluster with higher transverse energy. Clusters that are identified as local maxima give rise to L1 e/γ and τ candidates
3. **Particle Isolation:** For each cluster that is a local maximum, the pruned cluster deposits around it are summed creating isolation sums that can be used to determine if the cluster is isolated.
4. **Jet Reconstruction:** Overlap cleaned clusters are clustered together in jets corresponding in size to a cone of $\Delta R = 0.5$.
5. **Event Variable Estimation:** Clusters are summed together to form global event variables such as missing transverse energy ($\not{E}_T$) and sum of jets E_{T} (H_T).

In this chapter, the relevant algorithms for the performed analysis are discussed. These are related to Particle Finding, Filtering and Isolation.

2.1 Particle Cluster Finder

This first algorithm step aims at finding the energy deposits of single particles. It produces as basic building blocks all possible 2×2 clusters of adjacent trigger towers on the $\Delta(\eta) \times \Delta(\phi)$ lattice using a 2×2 sliding window.

This results in clusters that overlap by either one or two trigger towers. For each cluster, the E_T value of each trigger tower is compared to a selectable threshold and the E_T value is set to zero when the original value is below the threshold.

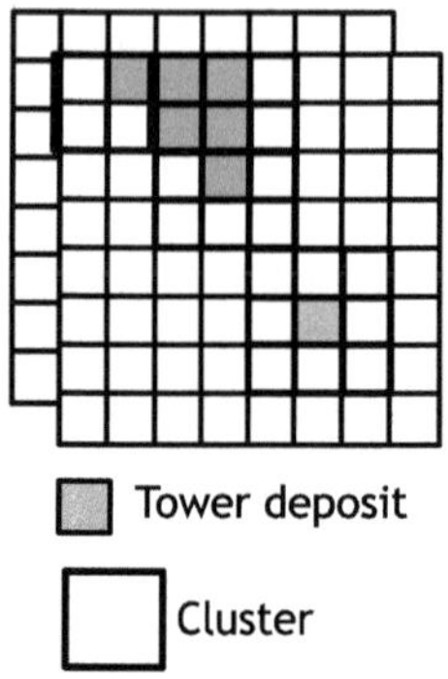

Fig. 1 Example of 2 × 2 clustering using a sliding window overlapping by one trigger tower. For each tower many clusters that overlap are created

This is done separately for the ECAL and the HCAL part. An energy deposit pattern as shown in Fig. 1 would yield 12 2 × 2 clusters with non-zero E_T. The E_T values above threshold are added up to provide $E_T^{\text{ECAL}} + E_T^{\text{HCAL}}$ per tower for each tower in the cluster. The OR of the fine-grain veto bits of the four towers is determined. The $\sum E_T^{\text{ECAL}}$ and the $\sum E_T^{\text{HCAL}}$ of the cluster is calculated as well and used as input to the electron photon identification module (EPIM). The EPIM applies calorimeter-based e/γ identification by comparing $\sum E_T^{\text{ECAL}} = E$ and $\sum E_T^{\text{HCAL}} = H$ for each cluster. Various suitable criteria are conceivable, for example $E/(E+H) >$ threshold, or $H/E <$ threshold, or thresholds that can be adjusted with the cluster E_T. An additional requirement is firmware stability. The implementation needs to be sufficiently flexible to allow changing the criterion without changing the firmware. In the simulation studies reported in this note, the EPIM criterion is a cut on $E/(E+H)$ that is relaxed with increasing E_T:

$$\frac{E}{E+H} > 0.95,\ E+H < 60\,\text{GeV} \tag{1}$$

$$\frac{E}{E+H} > 0.95 - 0.015(E+H-60),\ E+H > 60\,\text{GeV} \tag{2}$$

This optimizes the EPIM performance for electromagnetic objects with high energy that due to leakage of energy outside the ECAL, deposit an increasing amount of their energy in the HCAL.

2.2 Cluster Overlap Filter

For an energy deposit that is fully contained in 2×2 adjacent trigger towers the particle cluster finding step yields nine clusters with non-zero E_T. The cluster overlap filter determines which of these nine clusters contains the local maximum and prunes the remaining eight clusters such that their $\sum E_T$ no longer includes trigger towers belonging to the local maximum cluster.

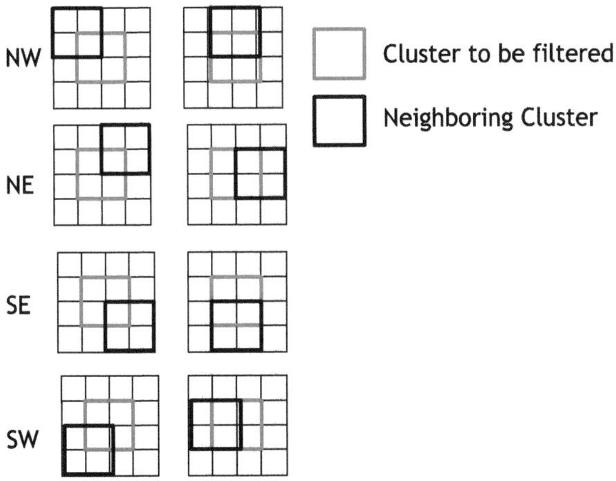

Fig. 2 Input to cluster overlap filtering and pruning algorithm

This is achieved as shown in Fig. 2. The E_T of the central cluster (C) is compared to each of the 8 neighboring clusters, located to the north-west (NW), north (N), etc. Whenever $E_T(C) < E_T(neighbor)$, the E_T of the one or two towers the two clusters have in common is pruned from the total E_T of the central cluster and the information that the central cluster is a pruned cluster is retained. A non-pruned cluster is by construction a local maximum. A cut on the $\sum E_T$ after pruning is applied and position information is calculated. One bit is retained to define if each cluster is pruned or if it is a local maximum and it is internally used to identify clusters are electrons or tau candidates. Clusters that have been pruned do not corresponmd to local maxima and they are not be assigned to particle candidates.

2.2.1 Cluster Position Determination

Determining the cluster position is done in parallel to the pruning step and hence yields meaningful results only for non-pruned clusters that correspond to particles. The position is calculated to half-tower precision, i.e. as one of the 16 points shown in Fig. 3. From the E_T values $E0$ to $E3$ of the four towers, the algorithm calculates

Fig. 3 Possible cases of the cluster position after cluster weighting

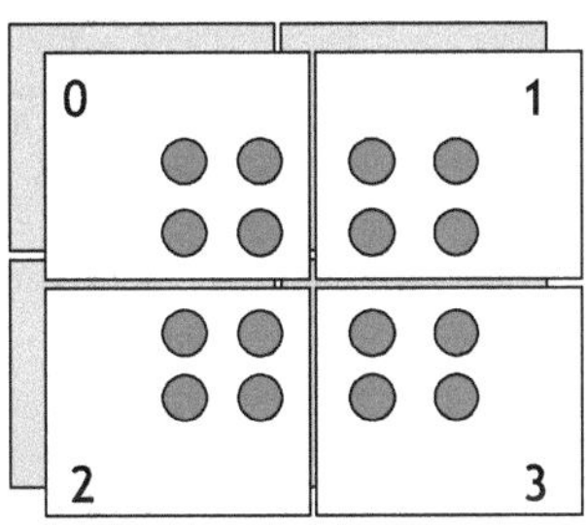

three E_T sums:

$$Ho = E1 + E3 - E0 - E2$$

$$Ve = E2 + E3 - E0 - E1$$

$$S = E0 + E1 + E2 + E3$$

The horizontal position is then given by $Hpos = Ho/Ve$, the vertical position by $Vpos = Ve/S$. The effect of cluster weighting is quantified in the simulation by comparing the generated particle position with the reconstructed particle position after matching the generated and reconstructed particle within a cone of $\Delta R = 0.5$. Figure 4 shows the position resolution in η and ϕ for the e/γ and tau objects respectively. The position accuracy achieved by the calorimetric algorithm is better than half of a trigger tower. The proposed segmentation is also much more precise than the current one.

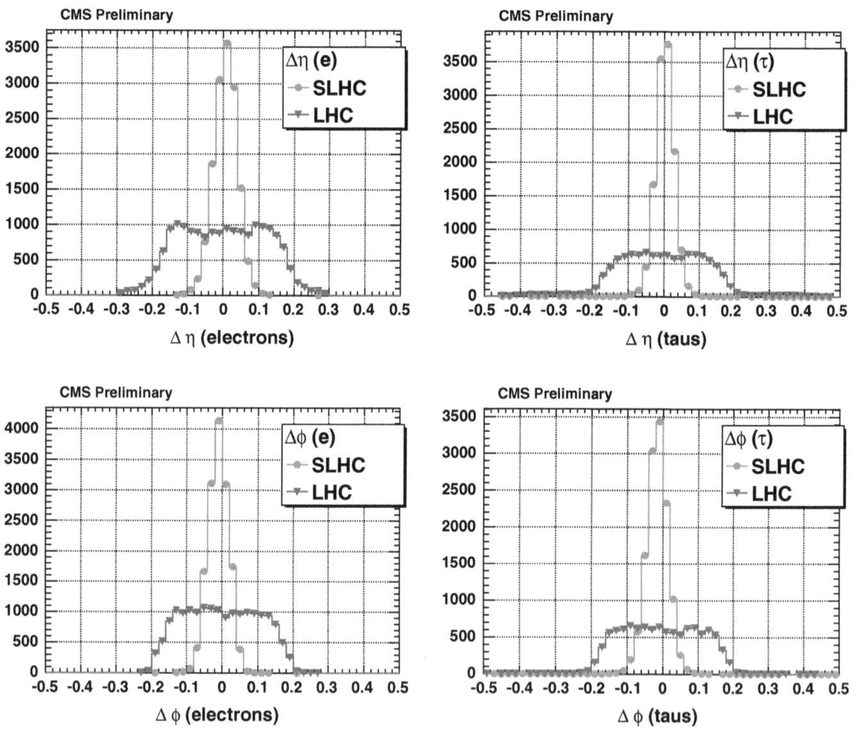

Fig. 4 Position resolution in η and ϕ for LHC and SLHC objects

2.3 Cluster Isolation

The algorithm considers a window of selectable size around a cluster flagged as the central one. It determines the number of 2×2 tower clusters with E_T above a selectable threshold within this window. This threshold can be chosen differently for electromagnetic and for τ objects. The size of the window can vary from 4×4 to 8×8 towers which corresponds to a cone size in $\eta - \phi$ space of between ~ 0.5 and ~ 1.0. The window size and the E_{T} of the central clusters may be used as input to a lookup table, providing the the possibility of employing a relative combined isolation criterion in the L1 trigger decision.

3 Performance Results

For studying the algorithm performance, a set of interesting physics samples was generated using the CMS Fast Simulation. Fast Simulation supports PU events and for this study, the average number of PU events generated is 25. Efficiency and resolution studies were conducted using $Z \rightarrow ee$ and $H \rightarrow \tau\tau$ samples, while the rate studies utilized a large sample of minimum bias events. The proposed SLHC algorithm is compared to the LHC algorithm currently used online in terms of the efficiency and the rate. For the efficiency and resolution studies the L1 objects are matched to the generator level objects. For the rate calculation, a bunch filling scheme with a fill fraction of 0.795 is used and the bunch crossing rate is 40 MHz. The rate is given by:

$$R = 40 \times 10^3 \times 0.795 \times \epsilon \left(E_{\mathrm{T}}^{thr}\right) \text{ kHz}, \tag{3}$$

$$R = 32 \times 10^3 \times \epsilon \left(E_{\mathrm{T}}^{thr}\right) \text{ kHz} \tag{4}$$

where ϵ is the probability for the trigger to fire at least one object of the specific type as a function of the trigger threshold E_{T}^{thre}. For these studies, this probability was defined as the number of events with an object at or above the threshold divided by the total number of events:

$$\epsilon\left(E_{\mathrm{T}}^{thr}\right) = \frac{(\text{Number of events with at least one object above } E_{\mathrm{T}}^{thr})}{(\text{Total number of events})} \tag{5}$$

3.1 Energy Calibration

The energy scale of the LHC and SLHC objects has been calibrated to have no systematic shift so that a fair comparison can be made. The applied calibration was a two-step process aimed to scale the energies of all relevant L1 objects to the energy

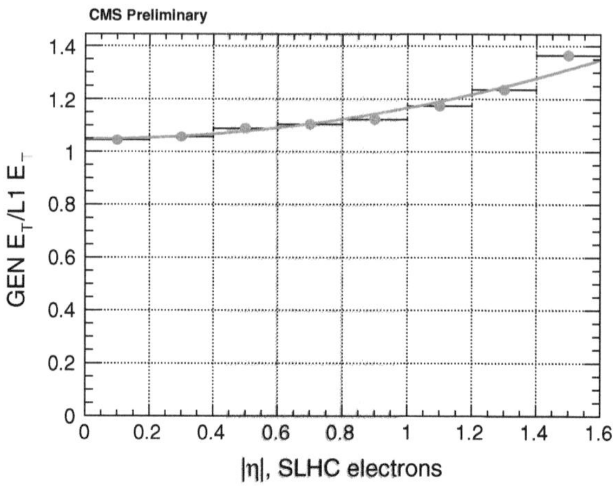

Fig. 5 Functional fit to the ratio GEN E_T / L1 E_T for the SLHC in the region $0 \leq |\eta| \leq 1.6$

of their matched generator-level objects. Electron and tau objects were calibrated using $Z \rightarrow ee$ and $H \rightarrow \tau\tau$ samples, respectively. A functional form was fit to the ratio (GEN E_T / L1 E_T) as a function of $|\eta|$ for all matched L1 objects with $E_T > 5$GeV. Second, a bin-by-bin correction factor was implemented so that the average ratio was 1 for the triggered objects in each $|\eta|$ bin. The transverse energy for each reconstructed then is scaled by an amount specified by both the functional form and its $|\eta|$ bin: This calibration curve is applied to all objects, and the average ratio GEN E_T / L1 E_T is obtained for each $|\eta|$ bin for those objects which pass the trigger threshold (Fig. 5).

3.2 Efficiency and Rate

The SLHC algorithm is compared to the current algorithm for isolated and non-isolated objects. The LHC algorithm for taus uses only isolated objects so isolated LHC taus are compared to both isolated and non-isolated SLHC taus.

Figure 6 shows the efficiency and rate for non isolated objects. In both e/γ and tau cases, the SLHC algorithm gives better rejection for higher efficiency. In the tau case the improvement is much more significant, showing that even non-isolated tau objects with SLHC algorithm provide much higher efficiency and much lower rate than isolated objects with the current LHC algorithm.

Figure 7 shows the efficiency and rate for isolated objects. The scope of the study is to compare the e/γ and tau rates for a working point that provides similar efficiency in these isolated cases by adjusting the parameters governing the SLHC algorithm's performance. The rate reduction of the SLHC algorithm for the same efficiency as the old algorithm is more than a factor of five for isolated taus and about a factor of three

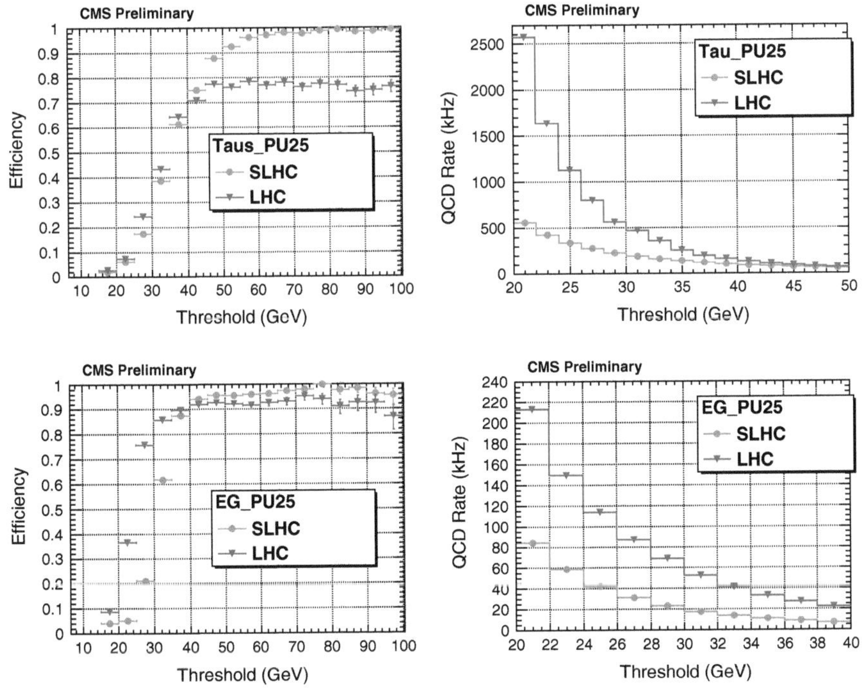

Fig. 6 Efficiency and rates for non-isolated objects

for isolated e/γ objects. Finally the most complicated algorithms were converted to firmware and simulation was performed on real FPGAs. The studies showed that the nature of the implementation makes those algorithms really applicable for hardware applications. In addition the fraction of the FPGA logic consumed by the most complicated algorithms was small [1] allowing for a 17×17 lattice size in a Xilinx Virtex-5 FPGA.

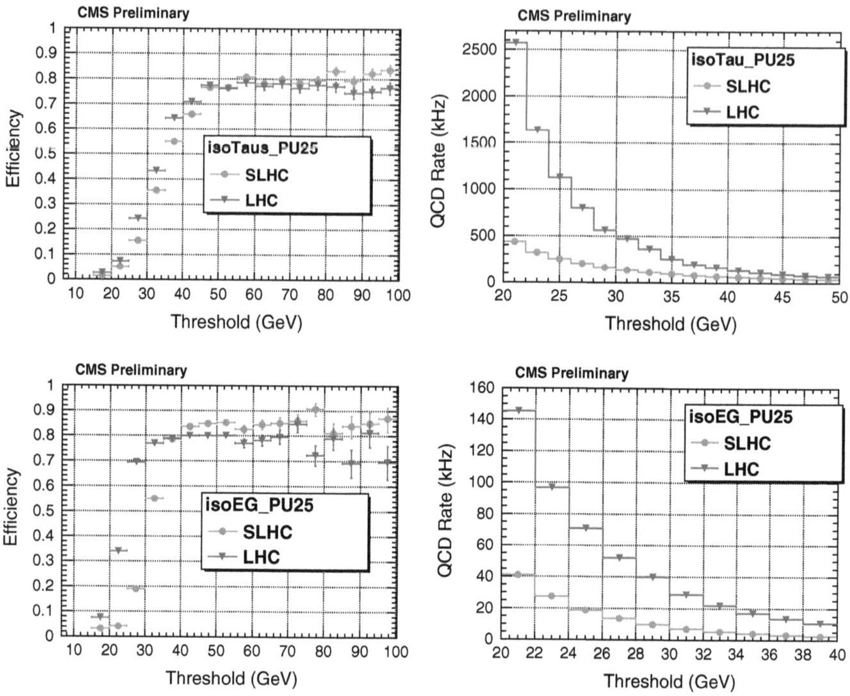

Fig. 7 Efficiency and rates for isolated objects

Reference

1. A. Gregerson, A. Farmahini-Farahani, B. Buchli, S. Naumov, M. Bachtis, K. Compton, M. Schulte, W.H. Smith, S. Dasu, FPGA Design Analysis of the Clustering Algorithm for the CERN Large Hadron Collider. In Field Programmable Custom Computing Machines, 2009. FCCM '09. 17th IEEE Symposium on, April 2009, pp. 19–26

MIX
Papier aus verantwortungsvollen Quellen
Paper from responsible sources
FSC® C105338

If you have any concerns about our products,
you can contact us on
ProductSafety@springernature.com

In case Publisher is established outside the EU,
the EU authorized representative is:
Springer Nature Customer Service Center GmbH
Europaplatz 3, 69115 Heidelberg, Germany

Printed by Libri Plureos GmbH
in Hamburg, Germany